INTERNATIONAL UNION OF CRYSTALLOGRAPHY TEXTS ON CRYSTALLOGRAPHY

IUCr Monographs on Crystallography

IUCr Texts on Crystallography

IUCr Crystallographic Symposia

Crystal Structure Analysis
Principles and Practice

William Clegg

Department of Chemistry, University of Newcastle upon Tyne

Alexander J. Blake

School of Chemistry, University of Nottingham

Robert O. Gould

Structural Biochemistry Unit, University of Edinburgh

Peter Main

Department of Physics, University of York

Edited by
William Clegg

INTERNATIONAL UNION OF CRYSTALLOGRAPHY

OXFORD
UNIVERSITY PRESS

*This book has been printed digitally and produced in a standard specification
in order to ensure its continuing availability*

OXFORD
UNIVERSITY PRESS

Great Clarendon Street, Oxford OX2 6DP

Oxford University Press is a department of the University of Oxford.
It furthers the University's objective of excellence in research, scholarship,
and education by publishing worldwide in

Oxford New York

Auckland   Cape Town Dar es Salaam   Hong Kong   Karachi
Kuala Lumpur Madrid Melbourne Mexico City   Nairobi
New Delhi Shanghai   Taipei Toronto
With offices in
Argentina Austria Brazil Chile Czech Republic France Greece
Guatemala Hungary Italy Japan South Korea Poland Portugal
Singapore Switzerland Thailand Turkey Ukraine Vietnam

Published in the United States
by Oxford University Press Inc., New York

© William Clegg, Alexander J. Blake, Robert O. Gould, and Peter Main, 2001

The moral rights of the author have been asserted

Database right Oxford University Press (maker)

Reprinted 2006

ISBN 0-19-850618-X

Preface

The material in this book is derived from an Intensive Course in X-Ray Structure Analysis organised on behalf of the Chemical Crystallography Group of the British Crystallographic Association and held every two years since 1987. As with a crystal structure derived from X-ray diffraction data, the Course contents have been gradually refined over the years and reached a stage in 1999 where we considered they could be published, and hence made available to a far wider audience than can be accommodated on the Course itself. The authors were the principal lecturers on the Course in 1999 and have revised and expanded the material while converting the lecture notes into a book format. Any readers who participated in the 1999 Course will detect a number of changes, particularly in the inclusion of some material not covered in lectures, some updating and shift in balance, and differences of style made necessary by a non-interactive format.

Because of its origin, this book represents a snapshot of the Intensive Course, which continues to evolve, especially as the subject of chemical crystallography is currently undergoing more rapid changes than for many years, mainly due to the widespread availability of area detector technology and the exponential increase in computing power. Nevertheless, we believe it emphasises principles which will remain valid for a considerable time, and the particular application of those principles can be adapted to new developments for some time to come.

Since the book owes its origins to the Course, we wish to acknowledge our large debt to those who have dedicated much effort to the organisation of the Course since its inception; without them this book would never have existed, even as an idea. The first five Courses were held at the University of Aston, where the local organisers Phil Lowe and Carl Schwalbe set a gold standard of Course administration and smooth operation, establishing many of the enduring characteristics valued by participants ever since. Following the move to Trevelyan College at the University of Durham, Vanessa Hoy and then Claire Wilson have developed these firm foundations to even further heights of excellence. Throughout the Course's history Judith Howard has provided overall guidance and expertise, particularly in fund-raising, and has spared the Course lecturers much concern with the practicalities of maintaining and promoting the Course. Several organizations, including the EPSRC, IUCr and commercial sponsors, have been long-standing and generous supporters of the Course.

Many colleagues have made contributions to the Course over the years, in lectures and in the crucial group tutorial sessions: a book format can never reflect the intensive interaction and lively atmosphere. These and the social aspects of the Course are probably at least as important in the memories of participants as the formal lecture presentations.

Most of all we are indebted to David Watkin. The Course was his brainchild in the first place and he worked extremely hard to launch it and establish it as the enduring success that it has become. He gave up his involvement in the Course for a while, before it reached the stage of encapsulation in book form, but his influence is firmly stamped on the whole character of the Course and hence on this book. It is to him that we dedicate this work.

One aspect of the tutorial group sessions of the Course has been retained in modified form in the book. Most chapters include exercises, for which answers are provided in an appendix. Readers are encouraged to tackle the exercises at leisure and not consult the answers until they are satisfied with their own efforts. In the spirit of the tutorials, these exercises may also prove beneficial as a basis for group discussion.

In acting as overall editor for the conversion of the material from lecture notes to book format, Bill Clegg also wishes to thank the University of Canterbury, New Zealand, for the award of an Erskine Visiting Fellowship in 1999, during the tenure of which substantial amounts of the revision and editing were carried out.

June 2001

Sandy Blake, *University of Nottingham*
Bill Clegg, *University of Newcastle upon Tyne*
Bob Gould, *University of Edinburgh*
Peter Main, *University of York*

Acknowledgements

We are grateful to authors and publishers for their permission to reproduce some of the figures that appear in this book, as follows.

Figures 1.1, 1.8, 4.1, 18.1, and 18.2 from W. Clegg: *Crystal structure determination*. Oxford University Press, Oxford, 1998.

Figures 1.2 and 3.2 from C. Giaccovazzo, H. L. Monaco, D. Viterbo, F. Scordari, G. Gilli, G. Zanotti and M. Catti: *Fundamentals of crystallography*. Oxford University Press, Oxford, 1992.

Figure 1.9 from J. P. Glusker and K. N. Trueblood: *Crystal structure analysis: a primer, Second edition*. Oxford University Press, Oxford, 1985.

Figures 1.3 and 1.5 from G. Harburn, C. A. Taylor and T. R. Welberry: *Atlas of optical transforms*. G. Bell, London, 1975.

Figure 3.5 from Traidcraft plc, Gateshead, UK.

Figure 3.6 and part of Table 3.6 from *International tables for crystallography, Volume A*. Kluwer Academic Press, Dordrecht, The Netherlands. Copyright 1983, International Union of Crystallography.

Figure 13.4 from *International tables for X-ray crystallography, Volume IV*. The Kynoch Press, Birmingham, UK. Copyright 1974, International Union of Crystallography.

Figure 9.1, reprinted by permission from G. N. Ramachandran and R. Srinavasan, *Nature*, **190**, 161. Copyright 1961, Macmillan Magazines Ltd.

Figure 14.1, reprinted with permission from W. Clegg, N. Mohan, A. Müller, A. Neumann, W. Rittner and G. M. Sheldrick, *Inorganic Chemistry*, **19**, 2066. Copyright 1980, American Chemical Society.

Figure 14.3 from D. Zobel, P. Luger, W. Dreissig and T. Koritsanszky, *Acta Crystallographica*, **B48**, 837. Copyright 1992, International Union of Crystallography.

Contents

1

A basic introduction to
X-ray crystallography

The material in the book will assume some basic knowledge of crystal structure determination, and more elementary texts should be consulted as appropriate. As readers may be at very different levels of background understanding and experience, we set out here some of the fundamentals of the subject which will be developed subsequently. This may indicate where some revision of basic texts is advisable, and will also provide helpful reference material during the reading of the rest of the book. A collection of mathematical explanations and formulae, and a short crystallographic dictionary are provided as Appendix 1 and Appendix 2 respectively.

1.1   X-ray scattering from electrons

The scattering of X-rays from electrons is called Thomson scattering. It occurs because the electron oscillates in the electric field of the incoming X-ray beam and an oscillating electric charge radiates electromagnetic waves. Thus, X-rays are radiated from the electron at the same frequency as the primary beam. However, most electrons radiate π radians (180°) out of phase with the incoming beam, as shown by a mathematical model of the process. The motion of an electron is heavily damped when the X-ray frequency is close to the electron resonance frequency. This occurs near an absorption edge of the atom, changing the relative phase of the radiated X-rays to $\pi/2$ radians (90°) and giving rise to the phenomenon of anomalous scattering.

1.2   X-ray scattering from atoms

There is a path difference between X-rays scattered from different parts of the same atom, resulting in destructive interference which depends upon the scattering angle. This reduction in X-ray intensity scattered from an atom with increasing angle is described by the atomic scattering factor, illustrated in Fig. 1.1. The value of the scattering factor at zero scattering angle is equal to the number of electrons in the atom. The atomic scattering factors illustrated are for stationary atoms, but atoms are normally subject to thermal vibration. This movement is taken into account by a simple modification of the scattering factor: see 'displacement parameters' in the dictionary (Appendix 2).

If anomalous scattering takes place, the atomic scattering factor is altered to take this into account. This occurs when the X-ray frequency is close to the resonance frequency

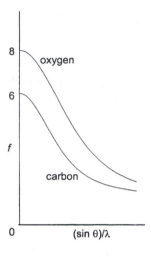

Fig. 1.1. The scattering factor of stationary carbon and oxygen atoms with no anomalous scattering effects.

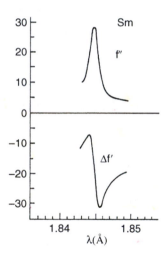

Fig. 1.2. Real and imaginary anomalous dispersion contributions for an Sm atom and a range of wavelengths.

of an electron. Only some of the electrons in the atom are affected and they will scatter the X-rays roughly $\pi/2$ radians out of phase with the incident beam. Electrons scattering exactly $\pi/2$ out of phase are represented mathematically by an imaginary component of the scattering factor and they cease to contribute to the real part. The exact phase change is very sensitive to the X-ray frequency. This is shown in Fig. 1.2, which displays the real and imaginary parts of the contribution to the atomic scattering factor of the anomalously scattering electrons as a function of wavelength for an atom of a particular element. The remaining electrons in the atom are unaffected by this variation in wavelength. Such information on atomic scattering factors is obtained from quantum mechanical calculations, and is available in standard reference tables; it is often built into crystallographic software packages for the most commonly used wavelengths.

1.3 X-ray scattering from the contents of a unit cell

X-rays scattered from each atom in the unit cell contribute to the overall scattering pattern. Since each atom acts as a source of scattered X-rays, the waves will add constructively or destructively in varying degrees depending upon the direction of the diffracted beam and the atomic positions. This gives a complicated scattering pattern whose amplitude and phase vary continuously, as can be seen in the two-dimensional optical analogue in Fig. 1.3.

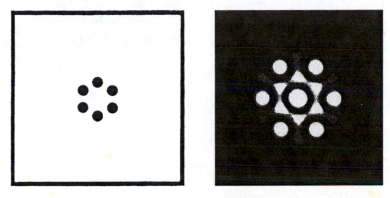

Fig. 1.3. A distribution of holes in an opaque sheet and their optical diffraction pattern.

1.4 The effects of the crystal lattice

The diffraction pattern of the pure crystal lattice is also a lattice, known as the reciprocal lattice. The name comes from the reciprocal relationship between the two lattices: large crystal lattice spacings result in small spacings in the reciprocal lattice and vice versa. The direct cell parameters are normally represented by $a, b, c, \alpha, \beta, \gamma$ and the reciprocal lattice parameters by $a^*, b^*, c^*, \alpha^*, \beta^*, \gamma^*$. The direction of a^* is perpendicular to the directions of b and c and its magnitude is reciprocal to the spacing of the lattice planes parallel to b and c. Similarly for b^* and c^*. A two-dimensional example of the relationship between the direct and reciprocal lattices is shown in Fig. 1.4.

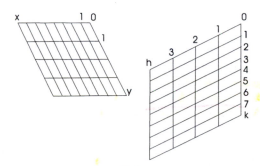

Fig. 1.4. Corresponding direct and reciprocal lattices in two dimensions.

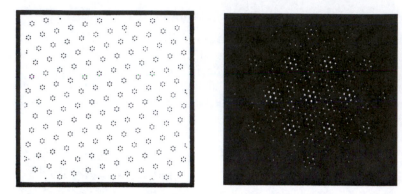

Fig. 1.5. The unit cell of Fig. 1.2 repeated on a lattice and its diffraction pattern.

1.5 X-ray scattering from the crystal

A combination (convolution; see Appendix 1 for mathematical treatment) of a single
unit cell with the crystal lattice gives the complete crystal. The X-ray diffraction pattern
is therefore given by the product of the scattering from the unit cell and the reciprocal
lattice, i.e. it is the scattering pattern of a single unit cell observed only at reciprocal
lattice points, as if the unit cell scattering pattern were being viewed through a mesh with
dimensions dictated by the lattice. This can be seen in Fig. 1.5 which shows the motif of
Fig. 1.3 repeated on a lattice and its corresponding diffraction pattern. The underlying
intensity is the same in both patterns. The positions of the reciprocal lattice points are
dictated by the crystal lattice; the intensity of the diffraction pattern at a reciprocal lattice
point is determined by the atomic arrangement within the unit cell.

1.6 The structure factor equation

There are many factors affecting the intensity of X-rays in the diffraction pattern. The
one which depends only upon the crystal structure is called the structure factor. It can
be expressed in terms of the contents of a single unit cell as:

$$F(hkl) = \sum_{j=1}^{N} f_j \exp[2\pi i(hx_j + ky_j + lz_j)]. \tag{1.1}$$

The position of the jth atom is given by the fractional coordinates (x_j, y_j, z_j), it has a
scattering factor of f_j and there are N atoms in the cell. Structure factors are measured
in number of electrons; they give a mathematical description of the diffraction pattern
such as that illustrated in Fig. 1.6. Each structure factor represents a diffracted beam
which has an amplitude, $|F(hkl)|$, and a relative phase $\phi(hkl)$. Mathematically, these
are combined as $|F(hkl)| \exp[i\phi(hkl)]$ and can be written as $F(hkl)$, which is a complex
number.

You may notice that the distribution of intensities in the diffraction pattern in
Fig. 1.6 is centrosymmetric. This is an illustration of Friedel's law, which states that

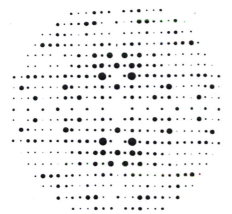

Fig. 1.6.   Part of the X-ray diffraction pattern of ammonium oxalate monohydrate.

$|F(hkl)| = |F(\overline{hkl})|$. The law follows from equation (1.1), which shows that $F(hkl)$ is the complex conjugate of $F(\overline{hkl})$, making the magnitudes equal and relating the phases as $\phi(hkl) = -\phi(\overline{hkl})$. This is no longer true when the atomic scattering factor f_j is also complex. Changing the signs of the diffraction indices does not produce the complex conjugate of f_j, so Friedel's law is not obeyed when there is anomalous scattering. However, the effect is phase dependent and for centrosymmetric structures, where all the phases are 0 or π radians, the magnitudes of $F(hkl)$ and $F(\overline{hkl})$ are always changed by the same amount.

The experimental measurements consist of the intensity of each beam and its position in the diffraction pattern. After suitable correction factors are applied, the quantities recorded are $h, k, l, |F(hkl)|$ (or F^2 instead of F), usually together with an estimated uncertainty for $|F|$ or F^2.

1.7   The electron density equation

An image of the crystal structure can be calculated from the X-ray diffraction pattern. Since it is the electrons which scatter the X-rays, it is the electrons which we see in the image, giving the value of the electron density at every point in a single unit cell of the crystal. The units of density are number of electrons per cubic angstrom unit, $e\,\text{Å}^{-3}$. The electron density is expressed in terms of the structure factors as:

$$\rho(xyz) = \frac{1}{V}\sum_{hkl} F(hkl)\exp[-2\pi i(hx + ky + lz)] \qquad (1.2)$$

where the summation is over all the structure factors $F(hkl)$ and V is the volume of the unit cell. Note that the structure factors include the phases $\phi(hkl)$ and not just the experimentally measured amplitudes $|F(hkl)|$, and both are needed. Since the X-rays are diffracted from the whole crystal, the calculation yields the contents of the unit cell averaged over the whole crystal and not the contents of any individual cell (this is important in cases of structural disorder). In addition, because of the finite time it takes to

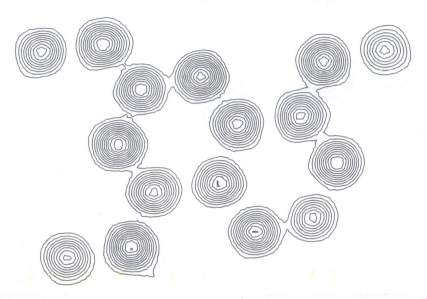

Fig. 1.7. A planar section of the three-dimensional electron density map of a molecule with many atoms in one plane.

perform the diffraction experiment, we see a time-averaged picture of the electrons. This results in a smeared-out image of each atom because of its thermal vibration, as seen in Fig. 1.7.

1.8   A mathematical relationship

Note the mathematical similarity between equations (1.1) and (1.2). Equation (1.1) transforms the electron density (in the form of atomic scattering factors, f_j) to the structure factors $F(hkl)$, while equation (1.2) transforms the structure factors back to the electron density. These are known as Fourier transforms, one equation performing the inverse transform of the other. This is a mathematical description of image formation by a lens. Light scattered by an object (Fourier transform) is collected by a lens and focused into an image (inverse transform). In the optical case, the (real) image is inverted and this is seen mathematically by the appearance of the negative sign in the exponent of equation (1.2).

1.9   Bragg's law

We cannot go far into X-ray diffraction without mentioning Bragg's law. This gives the geometrical conditions under which a diffracted beam can be observed. Figure 1.8 shows rays diffracted from lattice planes (sets of parallel planes passing through lattice points)

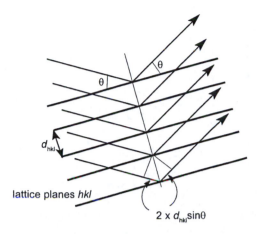

lattice planes *hkl*

$2 \times d_{hkl}\sin\theta$

Fig. 1.8.   Diffraction of X-rays from crystal lattice planes illustrating Bragg's law.

and, to get constructive interference, the path difference must be a whole number of wavelengths. This leads to Bragg's law which is expressed as:

$$2d \sin \theta = n\lambda \qquad (1.3)$$

where θ is known as the Bragg angle, λ is the wavelength of the X-rays and d is the spacing between adjacent planes in the parallel set. Figure 1.8 suggests the rays are reflected from the crystal planes. They are not—it is strictly diffraction—but reflection is mathematically equivalent in this context and the name X-ray reflection has stayed with us since Bragg first used it. The value of n in Bragg's law can always be taken as unity, since any multiples of the wavelength can be accounted for in the diffraction indices h, k, l of any particular reflection. For example, $n = 2$ for the planes h, k, l is equivalent to $n = 1$ for the planes $2h, 2k, 2l$, which have half the spacing.

1.10   Resolution

In X-ray crystallography we have effectively a microscope which gives images of crystal structures, although its realization is different from an ordinary optical microscope. What is the resolution of the image and what is its magnification? By convention, the resolution is given by the minimum value of d which appears in Bragg's law. This will correspond to the maximum value of θ for the set of diffraction data. With Mo Kα radiation and all data collected to a maximum θ of 25°, Bragg's law gives $2\sin(25°)/0.71 = 1/d_{min}$, producing a resolution (d_{min}) of 0.84 Å. The maximum possible resolution is $\lambda/2$, which occurs when $\sin\theta_{max} = 1$. For Cu Kα radiation this will be 0.77 Å, similar to the resolution obtained with Mo Kα for $\theta_{max} = 25°$. Figure 1.9 shows the effect on the electron density of limiting the extent of the diffraction pattern used to produce it.

If an electron density map is displayed on a scale of 1 cm/Å, this corresponds to a magnification of 10^8. You should be impressed by this very large number.

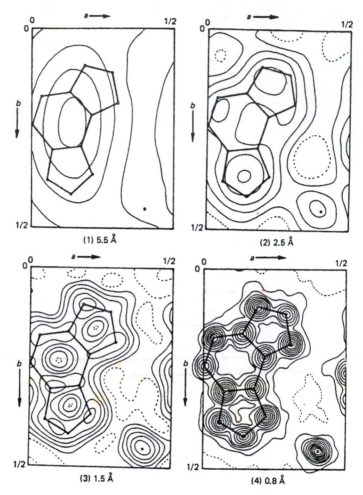

Fig. 1.9. The electron density calculated from a diffraction pattern of limited extent. The value of d_{min} is given in each case.

1.11   The phase problem

The measured X-ray intensities yield only the structure factor amplitudes and not their phases. The calculation of the electron density, therefore, cannot be performed directly from experimental measurements and the phases must be obtained by other means. Hence the so-called phase problem. Methods of overcoming the phase problem include (i) Patterson search and interpretation techniques; (ii) direct methods; (iii) use of anomalous dispersion; (iv) isomorphous replacement; (v) molecular replacement. Methods (i) (particularly the interpretation of a Patterson map when some heavy atoms are present) and (ii) are the most important in chemical crystallography; the others feature in macromolecular crystallography.

1.12   A flowchart for crystal structure determination

Figure 1.10 shows a simple representation of the main steps involved in determining a crystal structure. The various steps are treated in subsequent chapters.

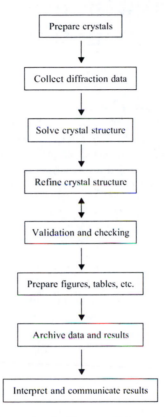

Fig. 1.10.   The steps involved in a crystal structure determination.

2

Crystal growth, evaluation and mounting

Whether growing crystals or **giving** advice to someone else trying to do so, it is vital to remember that the quality of **the** crystal from which diffraction data are acquired is generally the main determinant **of the** final quality of the structure. The effects of a sub-optimal crystal will propagate **through** data collection, structure solution and refinement to affect the quality of the final structure, in which unsatisfactorily high uncertainties may limit useful comparison and discussion. It may be difficult or impossible to get such a structure published.

2.1 Crystal growth

The term recrystallization has two related meanings, but there are crucial differences between these. In the synthesis and purification of compounds, the aim is to maximize purity and yield, although these can be mutually exclusive. The material is often precipitated very rapidly ($\sim$1 second), resulting in microcrystalline or nearly amorphous products that are useless for conventional single crystal work. For diffraction work, the object is to obtain a small number (one might be sufficient) of relatively large ($\sim$0.1–0.4 mm) single crystals. As long as this is achieved, yield is irrelevant and purity is likely to be enhanced. To this end crystals should be grown *slowly*, taking from minutes to months depending on the system. To understand why this is important, visualize the process of growth at a crystal surface. The greater the rate at which molecules arrive at the surface, the less time they have to orient themselves in relation to molecules already there; random accretion is more likely, leading to crystals which are twinned or disordered. Suitable growth conditions include the absence of dust and vibration; if these are present they can lead to crystals which are small or non-singular.

2.2 Survey of methods

2.2.1 Solution methods

These are by far the most flexible and widely used. They are suitable for use with molecular compounds that are the subject of most crystal structure determinations. The use of solvents means that crystals can grow separately from each other. It is therefore important not to let a solution dry out, as crystals could become encrusted and may not remain single. When choosing solvents, remember the general rule that 'like dissolves like': look for a solvent that is similar to the compound (in terms of polarity, functional groups, etc.)

and an anti-solvent (precipitant) that is dissimilar to it in order to reduce its solubility (see below). Information about solvents is available from several sources (e.g. *CRC Handbook of Physics and Chemistry* [1], chromatographic elution data), and the process of synthesizing and purifying a compound will often confer a knowledge of suitable solvents. Mixing solvents allows manipulation of solubility; a mixture of solvent A (in which a compound is too soluble) and anti-solvent B (in which it is not sufficiently soluble) may be more useful than either alone. If crystals grown from one solvent are poor, try different solvents or mixtures of solvents. Solution methods can be extremely flexible; a number of crystallizations, differing in the proportions of solvents A and B used, can be set up to run in parallel. If a particular range of proportions appears to be more successful in producing crystals, it can be investigated more closely by decreasing the difference between successive mixtures of A and B.

It is important that any vessels used for crystal growth should be free of contaminants. Older containers tend to have a large number of scratches and other surface defects, providing multiple nucleation points and tending to give large numbers of small crystals. Two factors which favour the formation of twinned crystals are the presence of impurities and uneven thermal gradients. Conversely, if the inner surface of a container is too smooth, this may inhibit crystallization. If this appears to be the case, gently scratching the surface with a metal spatula a few times may be effective. Some of the possible variations are described briefly below, and in almost all cases the methods described can be adapted to accommodate air sensitivity.

Cooling. Either make up a hot, nearly saturated solution and allow it to cool slowly towards room temperature, *or* make up such a solution at room temperature and cool it slowly in a refrigerator or freezer. The cooling rate can be reduced by exploiting the fact that the larger and more massive an object is, the longer it will take to lose heat. Thus, a hot solution in a large vessel (or in a small vessel within a larger one) will cool relatively slowly (Fig. 2.1). Similarly, a sample tube containing a solution will cool at a slower rate if it is contained in a metal block that was originally at room temperature, or if surrounded by an effective layer of insulation. Cooling methods are based on the generally valid assumption that solubility decreases with temperature. There are rare exceptions to this (e.g. Na_2SO_4 in water) and some solubilities rise so rapidly with temperature that it can

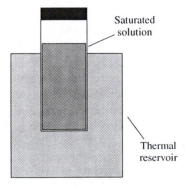

Saturated solution

Thermal reservoir

Fig. 2.1. Cooling.

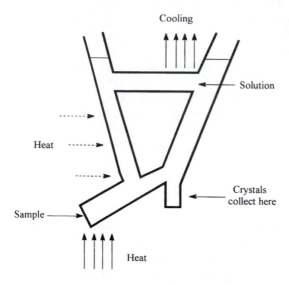

Fig. 2.2. Convection.

be difficult to control crystallization (e.g. of KNO_3 from water). However, it is usually possible to find a combination of solute and solvent where solubility varies slowly and controllably with temperature.

Convection. The aim here is to establish a temperature gradient across the solution, so that material dissolves in the warmer area and deposits in the colder. This gradient can be established by various means, for example (a) allow sunlight to shine on one part of the vessel; (b) put one part of the vessel against a cooler surface, such as a window at night; (c) construct an apparatus with low-power electrical heating elements in some sections (Fig. 2.2). A smooth concentration gradient will give the best results.

Concentration. If the volume of a solution is reduced, for example by evaporation of a volatile solvent, the concentration of the solute will rise until it begins to crystallize. When using mixed solvents the poorer solvent should be the less volatile so that the solubility of the solute decreases upon evaporation. The rate of evaporation can be controlled in various ways, for example by altering the temperature of the sample or by adjusting the size of the aperture through which the solvent vapour can escape. As solvents are frequently flammable or irritant, it is important to work on the smallest scale possible and ensure than any vapour released from the solution is safely dealt with. Avoid obvious hazards such as those that will arise if large volumes of diethyl ether or other highly volatile solvents are allowed to evaporate in a closed container such as a refrigerator.

As noted above, it is highly undesirable to let a solution evaporate to dryness, as this will allow otherwise suitable crystals to become encrusted, grow into an aggregate, or be contaminated by impurities. The crystals may be degraded by loss of solvent of crystallization, especially if chlorocarbon solvents such as dichloromethane have been used. It may prove impossible to identify good crystals even if these are present, and extracting them undamaged from a mass of material may prove difficult or impossible.

Sealed NMR tubes which have been forgotten at the back of a fume cupboard or refrigerator for weeks or months seem to be a fruitful source of good quality crystals; there is in fact slow evaporation of solvent, and crystals are able to grow undisturbed. As long as the NMR tube is clean and relatively unscratched, the smooth inner surface and narrow bore provide an excellent environment for crystal growth.

One method that combines variation of both concentration and temperature and is especially useful for sparingly soluble compounds is Soxhlet extraction. The recycling of the solvent is the key factor here and crystals can even appear in the refluxing solvent. Failing this, they normally appear after slow cooling of the solution. There are several other methods available to control concentration. One of these is based on osmosis, where the solvent passes through a semi-permeable membrane into a concentrated solution of an inert species. The resulting increase in the concentration of the solute may lead to crystal formation.

Solvent diffusion. This method is based on the fact that a compound will dissolve well in certain solvents ('good' solvents) but not in others ('poor' solvents or anti-solvents): these must be co-miscible. Dissolve the compound in the 'good' solvent and place this solution in a narrow tube. Using a syringe fitted with a fine needle, very slowly inject the neat anti-solvent. If it is lighter than the solution, layer it on top; if it is denser, inject it slowly into the bottom of the tube to form a layer under the solution. Injecting the solvent is better than running it down the side of the tube. If the tube is protected from vibration, these layers will mix slowly and crystals will grow at the interface (Fig. 2.3). If necessary, cooling of the tube can be used both to lower the rate of diffusion and to reduce the solubility.

Vapour diffusion. This method is also called isothermal distillation. The anti-solvent diffuses through the vapour phase into a solution of the compound in the 'good' solvent, thereby reducing the solubility (Fig. 2.4). The advantages of this method include the relatively slow rate of diffusion, its controllability and its adaptability, for example in combination with Schlenk techniques to grow crystals of air-sensitive samples. It is

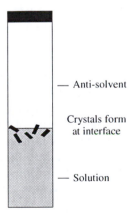

— Anti-solvent

Crystals form
at interface

— Solution

Fig. 2.3.   Solvent diffusion.

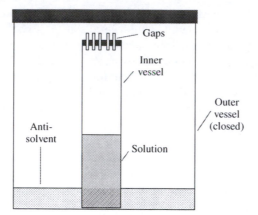

Fig. 2.4. Vapour
diffusion.

usually worth trying vapour diffusion, as it frequently succeeds where other methods
have failed. A variant on this is the hanging drop method, principally used for the growth
of crystals of proteins and other macromolecules, where the precipitant sits in a well and
diffuses slowly into a drop of solution suspended on a glass slide covering the well.

Reactant diffusion. It is sometimes possible to combine synthesis and crystal growth.
In favourable cases crystals may simply drop out of the reaction mixture, but the rate
of many reactions means that crystals form rapidly and are therefore small and of low
quality. If the reaction rate can be controlled by slow addition of one of the reactants, this
offers one way to overcome the problem. The best control is often achieved by controlling
the rate at which reactant solutions mix, by interposing a semi-permeable barrier (e.g.
membrane, sinter or an inert liquid such as Nujol) or by the use of gel crystallization
(see below). Another variant involves placing a solid reactant at the bottom of a tube,
covering it with a solvent in which it is known to dissolve slowly, and carefully adding
an upper layer consisting of a solution of a second reactant. The additional time required
for the solid to dissolve reduces the rate at which reaction can occur.

Crystals of zeolites and of many other materials with network structures cannot be
recrystallized and therefore can only be obtained from the reaction mixture. Fine-tuning
of the reaction conditions and varying the proportions and concentrations of reactants
probably offer the only realistic ways to control crystal size and quality.

Crystallization from gels is an under-exploited technique for obtaining single crystals
of compounds of low solubility. Because the mixing of the solutions is dominated by
diffusion through a viscous medium, undesirable competing processes such as convection and sedimentation are minimized. It is therefore possible to establish laboratory
conditions for crystallization that closely approximate the microgravity of space. A typical arrangement is a U-tube half-filled with gel, with a solution of one reactant in the
top of one arm and a solution of another reactant in the other. As gels are generally
colourless it is much easier to detect and isolate crystals of a strongly coloured product.
There are various recipes for the preparation of gels (e.g. [2,3]), and it is possible to
treat gels with organic solvents to produce versions suitable for use with hydrophobic or
moisture-sensitive compounds.

Seed crystals. Sometimes crystallization of a compound gives crystals which, although
of good quality, are too small for structure analysis. A small number of these can be used
as seeds by placing them in a warm saturated solution of the compound and allowing
the solution to cool slowly. Crystal growth might then occur preferentially at the seed
to give a suitably large single crystal. A container free of contaminants and scratches is
strongly recommended here.

2.2.2 Sublimation

Sublimation is the direct conversion of a solid material to its gaseous state. It has been
harnessed to produce solvent-free crystals of electronic materials but it is applicable to
any solid with a significant vapour pressure at a temperature below its decomposition or
melting point. The basic experimental arrangement is simple (Fig. 2.5): a closed, usually
evacuated vessel in which the solid is heated (if necessary) and a cold surface on which
crystals grow. If possible, avoid heating the solid, as lower sublimation temperatures
often lead to better crystals. If the solid sublimes too readily, the vessel can be cooled.
If a compound has a low vapour pressure, sublimation can be enhanced by evacuating
the vessel or by using a cold finger containing acetone/dry ice ($-78°C$) rather than cold
water ($5-10°C$).

2.2.3 Fluid phase growth

It is possible to grow crystals directly from liquids or gases, often by employing *in situ*
techniques. Fluid phase methods encompass both high-temperature growth from melts
and low-temperature growth from compounds that melt below ambient temperature.

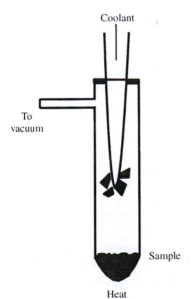

Fig. 2.5. Sublimation.

High-temperature methods (zone refining, etc.) are used widely in the purification and growth of crystals of semiconductors and other electronic materials, but are limited to compounds that melt without decomposition, thereby excluding many molecular compounds. Moreover, it is much more difficult to prevent unwanted phenomena such as twinning than with solution methods, and often impossible to separate overlapping or adjacent crystals. Liquids or gases must be contained, for example in a capillary tube. One consequence of this is that crystallization conditions must be controlled to give only one crystal in that part of the tube that will be within the X-ray beam. Once crystals have grown it is usually impossible to separate them physically. Unlike crystal growth from solution, there is essentially only one variable, namely the temperature of the sample. However, there are several ways to control the temperature and the method can be chosen to give coarse or fine control. A typical strategy for crystal growth involves the establishment and manipulation of a stable interface between liquid and solid phases. With air-stable compounds that crystallize in a fridge or freezer it is only necessary to keep them cold until they are transferred into the cold stream of the diffractometer's low-temperature device.

2.2.4   Solid-state synthesis

In favourable circumstances it may be possible to produce adequate single crystals, but microcrystalline samples are far more typical. For example, most high T_c superconductors do not give single crystals, and their structures have been determined using powder diffraction methods. As with the synthesis of zeolites from solution, variation of synthetic conditions is likely to be the only route to better single crystals.

2.2.5   General comments

The details of crystal growth are often poorly understood, especially for new compounds, and it is important not to be discouraged if initial attempts fail. For example, microcrystalline material is not immediately useful, but it does indicate that the compound is crystalline and that modification of the crystallization technique could result in larger crystals. It is always a good idea to try a range of techniques, keeping a detailed record of the exact conditions used and the results obtained. This not only allows identification of the most promising methods and conditions for the current sample but also means that in future there will be a database of procedures and their outcomes to consult. Crystal quality improves with experience, and early attempts often produce poor quality crystals. It is important to continue until it is clear that no further improvement is likely.

In some cases, regardless of the method employed, crystals either do not form or are unsuitable. At this stage, the best way to proceed may be to modify the compound. With ionic compounds it may be practical to change the counter-ion (e.g. BF_4^- for PF_6^-, or vice versa). With neutral compounds it may be a simple matter to change some chemically unimportant peripheral group. In one case altering a piperidine substituent to morpholine, which merely involved changing one remote CH_2 group for an oxygen atom, led to a spectacular improvement in crystal quality.

2.3 Sample evaluation

Once crystals have appeared, it is necessary to ascertain whether they are suitable for data collection. Some of the methods used are extremely rapid and can save large amounts of diffractometer time. During these procedures, take care to prevent damage to the crystals, for example by loss of solvent after removal from the mother liquor. If spare crystals are available, leave one or two exposed on a microscope slide and check them regularly for signs of deterioration, using microscopy as described below. It is vital to apply the tests outlined below *optimistically* so that only crystals that are clearly unsuitable are rejected. Any that give uncertain indications of their quality should be given the benefit of the doubt.

2.3.1 Microscopy

Visual examination under a microscope takes only a few seconds or minutes, yet can identify unsuitable crystals that might otherwise occupy hours on a camera or diffractometer. A microscope with a polarizing attachment, up to $\times 40$ magnification, a good depth of field, and a strong light source is required. Crystal examination consists of three steps.

Step one. With the analyser component of the polarizing attachment out (i.e. not in use), look at the crystals in normal light to determine if they are well-shaped. Reject crystals that are curved or otherwise deformed, have significant passengers which cannot be removed, or that show re-entrant angles. Be wary of rejecting crystals simply on the grounds that they are small, unless similarly sized crystals of the same type of compound have not previously been successful. For organic compounds containing no element heavier than oxygen, crystals smaller than $0.1 \times 0.1 \times 0.1 \text{ mm}^3$ seldom give good data with conventional laboratory instruments, although this size may be ideal for crystals of an osmium cluster compound.

Step two. With the analyser in, most crystals in a typical sample will transmit polarized light, unless they are in fact opaque. The exceptions are tetragonal, trigonal and hexagonal crystals viewed along their unique c axis, and cubic crystals viewed in any orientation. Tetragonal, trigonal or hexagonal crystals transmit polarized light when viewed along other directions, but cubic crystals cannot be distinguished from amorphous materials such as glass by this method. Fortunately, these crystal systems together account for less than 5% of molecular crystals.

Step three. If a crystal transmits polarized light, turn the microscope stage until the crystal turns dark (extinguishes), then light again, a phenomenon that will occur every $90°$. This extinction is the best optical indication of crystal quality, and it should be complete throughout the crystal and be relatively sharp ($\sim 1°$). Any crystal that does not extinguish completely is not single and can be rejected immediately. Lack of sharpness may indicate a large mosaic spread within the crystal. A crystal that never extinguishes is almost certainly an aggregate of smaller crystals. When examining a batch of crystals, establish both the general quality of the sample and whether there are individual crystals suitable for further study.

2.3.2 X-ray photography

An oscillation photograph taken using a Polaroid cassette can be obtained within 5–10 minutes, while a set of exposures from a precession or Weissenberg camera can require one or two days. As well as giving information on crystal quality, photographs can be used to establish unit cell dimensions and diffraction symmetry. Photography declined in popularity as data collection using four-circle instruments advanced, in part because it is often quicker to record a full data set than to obtain a set of photographs. However, it is worth remembering that photography gives a better view of the reciprocal lattice than can be obtained from the list of reflections output by a four-circle diffractometer, and can record any diffraction occurring at other than the expected positions. When screening crystals of dubious quality on a four-circle diffractometer, there is probably no faster method than Polaroid photography.

2.3.3 Diffractometry

The ultimate test of a crystal is how it behaves on the diffractometer. Reflections must possess sufficient intensity, be well-shaped (not split or excessively broadened), and index to give a sensible unit cell. Area detector instruments combine some of the best features of photographs and electronic counters, and some can establish the quality of a crystal in seconds. It is worth bearing in mind that area detectors can often tolerate lower quality crystals than four-circle instruments, so that crystals which would be rejected for data collection on a four-circle may be viable when using an area detector.

2.4 Crystal mounting

2.4.1 Standard procedures

For crystals that are stable to ambient conditions of air, moisture and light, the requirements of mounting are simple. The crystal is fixed securely with a reliable adhesive (e.g. epoxy resin) onto a glass or quartz fibre that is in turn glued into a 'pip' which fits into the well at the top of the goniometer head. The aim is to ensure that the crystal does not move with respect to this head. This means rejecting adhesives that do not set firmly (e.g. vaseline or Evo-Stik) or mounting media that are not rigid (e.g. plasticine, Blu-Tak or picene wax). On some diffractometers, crystals are spun at up to 4000° per minute, and an insecure mounting will lead to serious problems of crystal movement. A suitable fibre (e.g. of Pyrex glass) is just thick enough to support the crystal at a distance of about 5 mm above the pip. Fibres that are too thick add unnecessarily to errors via absorption and background effects, while those that are too thin can allow the crystal to vibrate, especially if it is being cooled in a stream of cold gas. For normal-sized crystals, the fibre should be thinner than the crystal. For small or thin crystals, use a 'two-stage' fibre, which consists of a glass fibre onto which is glued approximately 1 mm of glass wool to which the crystal is attached. The fibre confers stability, while the short length of glass wool reduces the amount of glass in the X-ray beam.

An outline of the basic procedure for mounting a crystal is as follows. First mix the epoxy resin, which will typically become tacky within five minutes and thereafter remain usable for a further five. Place the tip of the fibre into the resin and use the microscope to check that it has become coated. Ideally, the aim is to attach the tip of the fibre to the side of the crystal, thereby minimizing the amount of glass in the X-ray beam. Establish the size of the crystals, cutting them to size with a scalpel or razor blade if necessary. When picking up a crystal there is a danger of gluing it onto the slide, but this can be easily avoided. Move the adhesive tipped fibre forwards until it makes contact with the side of the crystal, then continue moving the fibre forwards and upwards to lift the crystal clear of the slide. (With thin plates this procedure may not be possible. If there is no alternative to mounting a crystal with a fibre along an edge or across a face of the crystal, the fibre must be as thin as possible: a 'two-stage' fibre may be appropriate.) When picking up a crystal ensure (a) that no crystal axis aligns perfectly with the fibre (and therefore with the diffractometer ϕ axis), as this can enhance systematic errors due to Renninger (multiple reflection) and other effects, and (b) that the crystal height can be adjusted to bring it into the X-ray beam (it is frustrating to find later that this cannot be done because a fibre is too long or too short; on many instruments the X-ray beam passes 68 mm above the upper surface of the ϕ circle).

Instead of a simple fibre, some crystallographers prefer to mount the crystal on the end of an open capillary tube (less glass in the beam for the same diameter), on a number of short lengths of glass wool attached to a thicker fibre (for the same reason), or on quartz fibres (more rigid for a given diameter). Some of these methods are illustrated in Fig. 2.6.

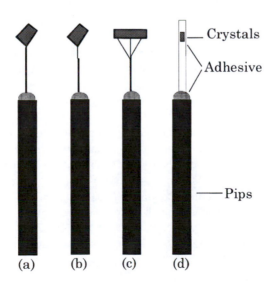

Fig. 2.6. Some methods for crystal mounting: (a) on a simple glass fibre; (b) on a 'two-stage' fibre; (c) on a fibre topped with several lengths of glass wool; (d) within a capillary tube.

2.4.2 Air-sensitive crystals

The traditional way to protect sensitive crystals is to seal them (using a flame or epoxy resin) into a capillary tube, usually made from Lindemann glass, which is composed of low molecular weight elements. This does, however, put a large volume of glass in the X-ray beam, so the tube wall should be as thin as practicable. The most sensitive crystals can be handled and encapsulated within a dry box. When planning a low-temperature data collection, ensure that the top end of the tube is well rounded and that there are only a few millimetres of glass above the crystal position, otherwise severe icing will result (alternatively, see the low-temperature methods described below). Crystals that desolvate may need either solvent vapour or mother liquor sealed into the tube with them. Unless crystals are mechanically robust, care must be taken when loading them into capillary tubes. With crystals that are both fragile and susceptible to solvent loss, a variant of a technique used by protein crystallographers may be helpful. Break the sealed end off a capillary tube and coat the first few millimetres of its inner surface with freshly mixed epoxy resin, place some crystals with their mother liquor in a well and isolate a good crystal, and bring the open end of the tube through the surface of the solution. It may take some practice, but capillary action should draw the crystal along with some mother liquor into the tube; the crystal will stick to one side of the tube, which can then be sealed at both ends.

Many crystals can be protected by coating them with materials such as nail varnish, superglue or epoxy resin. As long as the coating confers sufficient protection and does not react with the crystal, this can be a simple and effective solution to air-sensitivity that is applicable when cooling of the crystal is impossible. This situation can arise because a phase change is known or suspected to occur below ambient temperature, or because cooling causes an unacceptable degree of mechanical strain within the crystal.

A low-temperature device permits the use of an extremely flexible method for handling air-sensitive crystals. This involves transferring, examining and mounting the crystal under a suitably viscous inert oil. Upon cooling, the oil forms an impenetrable film around the crystal and also acts as an adhesive to attach the crystal firmly to the fibre. For crystals that do not survive room temperature, the technique can be combined with low-temperature handling, which normally involves passing a stream of cold nitrogen gas across the microscope stage. Various oils have been used, but perfluoropolyether oils have the advantages of inertness and immiscibility with solvents. The excellent Riedel de Haen RS3000 is no longer produced, but an alternative (PFO-XR75) is available from Lancaster Synthesis, although it is not quite so inert. An alternative method, popular with protein crystallographers and suitable for very thin crystals that are too fragile to be picked up on a fibre, is the solvent loop. A small loop of a fibre such as mohair or a single strand from dental floss is used to lift the crystal in a film of solvent or oil that is then flash cooled on the diffractometer to immobilize the crystal [4].

The final step is to attach the goniometer head to the ϕ circle of the diffractometer and optically adjust the crystal so that its centre does not move when ϕ and χ are rotated. Do not assume that the microscope cross-hairs represent the true centre. Centring is an iterative procedure. The following procedure is typical for Eulerian cradle instruments. First, with χ at $0°$, make sure the crystal is approximately central in X and Y by checking at $\phi = 0°, 90°, 180°$ and $270°$, then set the height Z approximately. Second, view the crystal at ϕ $0°$ and $180°$, then at $90°$ and $270°$. At any position the lateral offset must be the same as that seen $180°$ away in ϕ. Third, view the crystal at χ -90 and $+90°$. Z is correct if the offset is the same at both positions. Fourth, with χ at $0°$ re-check the crystal at ϕ $0°, 90°, 180°$ and $270°$. Repeat the third and fourth steps until convergence is achieved. The above procedure will require adaptation for different instrument geometries. For example, on an Enraf-Nonius CAD4 diffractometer, X and Y are checked at $\kappa = -60°$, and Z at $\kappa = \pm135°$.

References

[1] *CRC Handbook of Physics and Chemistry*, CRC Press, London; editions issued at frequent intervals.
[2] http://www.cryst.chem.uu.nl/lutz/growing/gel.html
[3] H. Arend and J. J. Connelly, *J. Cryst. Growth*, 1982, **56**, 642.
[4] E. F. Garman and T. R. Schneider, *J. Appl. Cryst.*, 1997, **30**, 211.

Some references on crystal growth and handling

H. E. Buckley, *Crystal Growth*, Wiley, London, 1951.
> Very detailed, good for background, and a source of alternative ideas for growing crystals. Strangely, there is almost no mention of sublimation.

P. M. Dryburgh, B. Cockayne and K. G. Barraclough (eds), *Advanced Crystal Growth*, Prentice Hall International (UK) Ltd, 1987.

P. G. Jones, Crystal growing, *Chem. Br.*, 1981, 222.
> This article also covers aspects of crystal evaluation and is highly recommended.

T. Köttke and D. Stalke, *J. Appl. Cryst.*, 1993, **26**, 615.
> The classic paper on the use of oil films for handling sensitive crystals. Excellent on practical aspects.

Look at the literature on related compounds. At the very least, the authors should have identified the solvent they used and the temperature at which crystals were grown.

Crystal growing hints and tips on the Web

http://laue.chem.ncsu.edu/web/GrowXtal.html
http://www.cryst.chem.uu.nl/lutz/growing/gel.html
http://www.cryst.chem.uu.nl/lutz/growing/reading.html
http://www.cryst.chem.uu.nl/growing.html

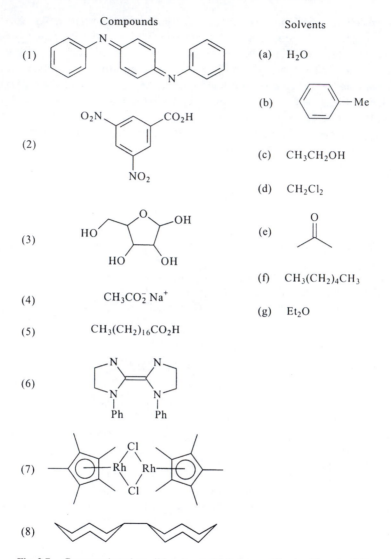

Fig. 2.7. Compounds and possible solvents for their crystallization (Exercise 2.1).

Exercises

2.1 Choosing suitable solvents. For each compound in the first column of Fig. 2.7, select from the second column a solvent in which it is likely to have good solubility.

2.2 Examination of crystals. In the following examples the behaviour of crystals examined using a polarizing microscope is described. In each case determine whether the crystals are likely to be suitable for structure determination, giving your reasons in each case.

(a) A sample of an organic compound containing no element heavier than nitrogen shows no crystals larger than $20 \times 30 \times 60 \, \mu m$.

(b) With crossed polars, crystals of a metal–organic compound remain dark at all angles regardless of their orientation.

(c) With crossed polars, crystals of another metal–organic compound transmit light at all angles (i.e. no extinction is ever observed).

(d) With crossed polars those crystals of an inorganic material which are showing square faces remain dark at all angles, while those showing rectangular faces exhibit extinction every 90°.

(e) Crystals of another sample show extinction every 90°, with the crystals remaining dark throughout a rotation of 6°.

(f) Although a crystal appears to consist of four components that all show extinction at different angles, the extinction is sharp in each case.

3

Symmetry and space group determination

3.1 Introduction

Symmetry is a feature of the world that some observers find fascinating and others merely annoying. Whether or not it is aesthetically appealing, it is certainly an organizing principle that can greatly simplify the analysis of many problems. Unfortunately, a trick that nature often plays is to make something *almost* symmetrical in an extremely tantalizing fashion. Crystallography is a subject that makes much use of symmetry; in fact, any understanding of crystals and crystal structure is almost impossible without some knowledge of the underlying symmetry of the crystals. The standard reference on crystallographic symmetry is the *International Tables for Crystallography, Volume A*. A very reasonably priced and strongly recommended introduction, with many of the most important features, is the special *Teaching Edition* of Volume A, which also covers problems in one and two dimensions.

Molecular materials normally have little molecular symmetry and only occasionally make use of it in their packing, but even so, the analysis of intermolecular interactions invariably makes use of symmetry in some form. More seriously, the collection of data, the solution of a structure and its refinement all make extensive use of crystallographic symmetry. Most readers will have had an introduction to molecular symmetry and the determination and use of molecular point groups. The possibilities for symmetry in an infinite array like a crystal differ in two ways from those of finite objects, like molecules. The good news is that the presence of a lattice restricts the possibilities of rotational symmetry to a total of five kinds, rather than the infinite possibilities for a molecule. The bad news is that symmetry involving *translation* is possible, since it is no longer necessary by repetition eventually to bring a point back to its starting position—an equivalent position in another unit cell will do as well. One result of these two conditions is that there are exactly 230 combinations of symmetry elements which can describe the symmetry of a crystal. These are called *space groups*, and our aim in this chapter is to see how the space group for a particular crystal may be determined. There is no unique way to proceed—a good deal of intuition is generally required, and the results of any automatic program should be carefully inspected! The steps generally (considered in greater detail later) are something like this:

1. For the first steps, it is wise, if in any doubt, to consider the *lowest* possible symmetry first; otherwise, only part of the diffraction data may be measured or analysed.
    (a) Derive any information possible from the morphology of the crystals or from knowledge that the crystals must be chiral.

(b) Determine the size and shape of the unit cell and, if possible, the density and composition of the crystals to determine how many formula units there are per unit cell.

(c) Determine the diffraction symmetry of the crystal and the unit cell centring (P, A, B, C, I, F or R).

2. Beyond this point, if in any doubt, start with the *highest* possible symmetry first and work downwards, only trying lower symmetry when forced to do so. Otherwise, effort may be wasted finding two or more 'independent' molecules which are actually related, and refining the resulting excess parameters.

(a) Determine any symmetry operations which involve translation and are indicated by special conditions for some classes of data ('systematic absences').

(b) Using the *International Tables*, check whether a space group has been uniquely determined.

(c) Examine the intensities statistically to see whether symmetry operations not involving translation are indicated.

3.2   Basic operations and point groups

As was mentioned above, the fact that the unit cell of a crystal must reproduce itself regularly on a lattice restricts the number of symmetry operations defining crystal morphology: in fact, there are only 10 operations to be considered. These are given here (Table 3.1), with their International (Hermann–Mauguin) and Schönflies notations. Note that in the International system, 'improper' rotations are combinations of rotations and inversions, while in the Schönflies system rotations and reflections are used.

The point groups (or classes) which can describe a crystal are all those which can be constructed using these and only these operations, and are thus 32 in number. Diffraction symmetry, in the absence of anomalous effects, is centrosymmetric, so the non-centrosymmetric crystallographic point groups may conveniently be classified together with the centrosymmetric one which describes the diffraction symmetry (Table 3.2). These 11 *diffraction* or *Laue groups* give the minimum symmetry information which may be determined from the diffraction pattern.

Lest this may seem rather daunting, it should be mentioned at this point that the various crystal systems are far from equally common, and the most common ones for molecular crystals are the least symmetric—those based exclusively on the three twofold operations 2, m and $\bar{1}$ (see Table 3.3).

Table 3.1   Crystallographic symmetry operations

Proper rotations		'Improper' rotations	
International	Schönflies	International	Schönflies
1	C_1 (identity)	$\bar{1}$	S_2, C_i or i (inversion)
2	C_2	$\bar{2}$ or m	S_1, C_s or σ (reflection)
3	C_3	$\bar{3}$	S_6
4	C_4	$\bar{4}$	S_4
6	C_6	$\bar{6}$	S_3 or C_{3h}

Table 3.2   Crystal systems, diffraction symmetries and point groups

System	Diffraction symmetry	Corresponding lower symmetries
Triclinic	$\bar{1}\,(C_i)$	$1\,(C_1)^*$
Monoclinic	$2/m\,(C_{2h})$	$m\,(C_s),\,2\,(C_2)^*$
Orthorhombic	$mmm\,(D_{2h})$	$mm\,(C_{2v}),\,222\,(D_2)^*$
Tetragonal	$4/m\,(C_{4h})$	$4\,(C_4)^*,\,\bar{4}\,(S_4)$
	$4/mmm\,(D_{4h})$	$4mm\,(C_{4v}),\,422\,(D_4)^*,\,\bar{4}m2\,(D_{2d})$
Trigonal	$\bar{3}\,(S_6)$	$3\,(C_3)^*$
	$\bar{3}m\,(D_{3d})$	$32\,(D_3)^*,\,3m\,(C_{3v})$
Hexagonal	$6/m\,(C_{6h})$	$6\,(C_6)^*,\,\bar{6}\,(C_{3h})$
	$6/mmm\,(D_{6h})$	$6mm\,(C_{6v}),\,622\,(D_6)^*,\,\bar{6}m2\,(D_{3h})$
Cubic	$m\bar{3}\,(T_h)$	$23\,(T)^*$
	$m\bar{3}m\,(O_h)$	$432\,(O)^*,\,\bar{4}3m\,(T_d)$

*Point groups possible for optically pure, chiral structures.

Table 3.3   Symmetry restrictions on unit cell parameters

System	Point group	a	b	c	α	β	γ	CSD*	ICSD*
Triclinic(6)	$\bar{1}$	given	given	given	given	given	given	(21.1)	(5.2)
Monoclinic(4)	$2/m$	given	given	given	$=90°$	given	$=90°$	(53.2)	(22.0)
Orthorhombic(3)	mmm	given	given	given	$=90°$	$=90°$	$=90°$	(20.8)	(24.1)
Tetragonal(2)	$4/mmm$	given	$=a$	given	$=90°$	$=90°$	$=90°$	(2.3)	(14.7)
Trigonal(2)	$\bar{3}m$	given	$=a$	given	$=90°$	$=90°$	$=120°$	(1.6)	(10.6)
Hexagonal(2)	$6/mmm$	given	$=a$	given	$=90°$	$=90°$	$=120°$	(0.5)	(7.2)
Cubic(1)	$m\bar{3}m$	given	$=a$	$=a$	$=90°$	$=90°$	$=90°$	(0.5)	(16.2)

All independent values are given; the number of these is in parentheses after the system name. The other values may be implied as above.

*Frequency (as percent) of 186 074 (largely molecular) entries in the Cambridge Structural Database (CSD) and 46 010 (largely non-molecular) entries in the Inorganic Crystal Structure Database (ICSD)—1999. Notice how different these lists are!

The point group of a crystal is the point group of the unit cell. The unit cell shape depends on this, along with the number of independent parameters required to define it. There are six distinct shapes of unit cells in terms of their symmetry requirements; these are usually referred to as the seven *crystal systems*, and are summarized in Table 3.3. It is important to realize that the shape of the unit cell is a result of the symmetry, *and not vice versa*.

3.3 External morphology

The external appearance of a crystal was the first indication of their atomic nature. Unfortunately, beautiful crystals of gem-like quality are the exception rather than the rule for chemical crystallographers. Nevertheless, even ordinary crystals will sometimes give a clue as to the underlying symmetry; the examples in the accompanying diagrams show what is possible in particularly favourable cases. A selection of these is shown in Fig. 3.1.

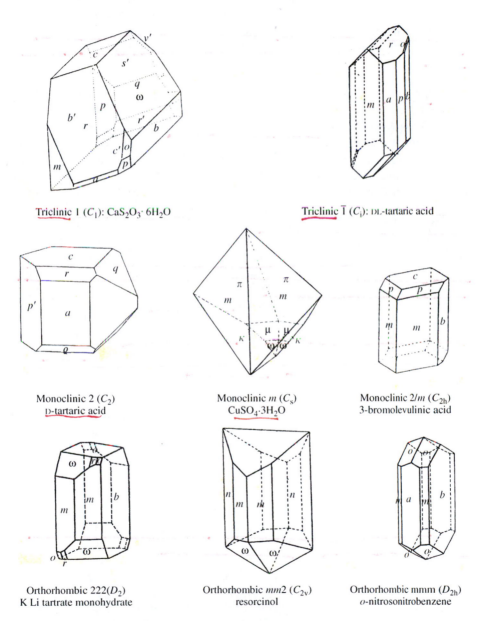

Triclinic 1 (C_1): $CaS_2O_3 \cdot 6H_2O$

Triclinic $\bar{1}$ (C_i): DL-tartaric acid

Monoclinic 2 (C_2)
D-tartaric acid

Monoclinic m (C_s)
$CuSO_4 \cdot 3H_2O$

Monoclinic 2/m (C_{2h})
3-bromolevulinic acid

Orthorhombic 222(D_2)
K Li tartrate monohydrate

Orthorhombic $mm2$ (C_{2v})
resorcinol

Orthorhombic mmm (D_{2h})
o-nitrosonitrobenzene

Fig. 3.1. Representative crystals in triclinic, monoclinic and orthorhombic point groups (*from Chemische Kristallographie*, by P. Groth, Wilhelm Engelmann, Leipzig, 1906–1921).

Probably the most striking feature which can be seen is polarity, which is present in the point groups 1, 2, *m* and *mm*2 and absent in the others shown here. Polarity indicates the absence of an inversion centre and of some other symmetry elements, such as three mutually perpendicular twofold axes (point group 222). Since crystals generally grow in an asymmetrical environment, it is dangerous to assume that crystals are polar on the grounds of morphology, unless it is clearly a property of several crystals in a sample. The polar point groups in Schönflies notation are those with symbols C_n or C_{nv}. In International notation, that corresponds to the point groups 1, 2, 3, 4, 6, *mm*2, 3*m*, 4*mm* and 6*mm*. Another property which can be deduced at the start is that any optically pure chiral compound must crystallize in a chiral point group: these point groups are also called 'holoaxial', because the only symmetry possible in them is pure rotation, and they are marked with an asterisk in Table 3.2.

Physical properties such as piezo-, pyro- and ferroelectric behaviour can, in principle, help to define possible crystal symmetry. In fact, few of these are of practical value, since they are not easy to measure and often produce negligible effects. Optical properties should, however, always be investigated. An inexpensive, low-power microscope with a polarizing attachment can often instantly indicate whether a crystal is single and give some information about its underlying symmetry. Amorphous material will have no effect on the plane of plane-polarized light. Thus, when an amorphous sample is viewed between crossed polars, it will always appear dark, like the field behind it. Under the same conditions, all triclinic, monoclinic and orthorhombic crystals (known as optically biaxial), viewed from any direction, will shift the plane of polarization in all orientations except two at right angles to one another. Thus, between crossed polars, they will appear bright except in these two positions, called extinction positions. The absence of a well-defined extinction position usually indicates a multiple crystal, unsuitable for diffraction. Tetragonal, trigonal or hexagonal crystals (called optically uniaxial) are similar, except when viewed along their unique axis. In this position they will appear isotropic, like amorphous material. Cubic crystals (optically isotropic) always appear like amorphous material. Fortunately, these usually have some well-defined faces, which indicate that they may well be worth testing for diffraction.

The contents of the unit cell should be considered for what they can indicate about symmetry. If the density of the crystal is measured, it is related to the unit cell volume by the relationship:

$$\rho = \frac{MZ}{N_A V}$$

where ρ is the density, M the mass of a mole, Z the number of formulae per unit cell, N_A is Avogadro's number and V the volume of the unit cell. If the density has not been measured, a rough value for Z may be estimated by dividing the cell volume (in Å^3) by 18 times the number of non-hydrogen atoms in the chemical formula. This remarkably simple rule is substantially correct for a wide range of organic and organometallic compounds. When the symmetry of the cell is being considered, Z may be used either to rule out some types of symmetry which are incompatible with the chemical formula or to question a chemical formula which is incompatible with the probable symmetry. For compounds like proteins, this approach must be used with some caution without a good estimate of the amount of solvent in the crystal.

3.4   Diffraction symmetry and the amount of independent data

With a serial diffractometer, the diffraction symmetry must be determined before data are collected, to enable data to be collected efficiently. With an area detector, this is less important. The actual amount of data is related to the volume of the unit cell and to the resolution of the data required. The total amount of data, not allowing for symmetry, is approximately $4.2V/d^3$, where V is the volume of the unit cell in Å^3 and d is the resolution of the data in Å. For $d = 0.84$ Å ($\theta \approx 67°$ with Cu radiation or $25°$ with Mo), a rough value is $7V$. This figure will be reduced by symmetry equivalences. Assuming that anomalous data are not required, all data related by the symmetry of the diffraction point group (Laue group) are equivalent and only one of a set needs in principle to be measured. When crystals are not centrosymmetric and there is significant anomalous scattering, $I(hkl) \neq I(\overline{hkl})$, so only data related by true point group symmetry are equivalent. Table 3.4 shows equivalences for each Laue group and those for the related chiral point group. Note that different relationships will apply for other point groups. General data are reflections with no special values (particularly zero) for any of the indices.

In most cases, not all data are 'general', e.g. in Laue group $2/m$, $h0l$ only has one other equivalent and not three. Thus the amount of data which needs to be measured will be somewhat more than the total amount divided by the general multiplicity.

A further consideration is the Bravais lattice of the crystal. The unit cell of a crystal may always be chosen as a primitive one, that is one in which no point within the cell is exactly like any other in orientation and environment. Thus a crystal with point group mmm (D_{2h}) may have a unit cell which is a rectangular prism or a rhombic prism, in which two of the axes are equal but are not parallel to the twofold axes of the point group. If the cell of such a crystal is chosen as the much more convenient rectangular prism, the cell will have twice the size of the primitive one. It is said to have two 'lattice points' per cell, as any point in the cell will have one other that is exactly identical to it in orientation and environment. Artificially doubling the size of the unit cell clearly cannot increase the number of diffraction data, so half of the data predicted for the new cell must be absent. There are several types of such cells, and their properties are given in Table 3.5. The combination of the crystal system and the lattice centring is called the Bravais lattice of the crystal. There are 14 distinct Bravais lattices, which are summarized later in Table 3.11.

3.5   Internal symmetry and translational symmetry operations

Unlike finite molecules, crystals are essentially infinite in all three directions, and so have other possibilities for symmetry. For example, a rotation of $180°$ applied twice returns the system to its original location. Another possibility is a translation of half a unit cell along the axis with each rotation. Two such rotations will return the system not to its original position but to an equivalent one, one unit cell removed from the first. Such operations, or 'screw axes' are possible for two-, four- and sixfold axes, and are called 2_1, 4_2 and 6_3 screw axes; the fraction of the unit cell travelled in each operation

Table 3.4 Equivalent data for diffraction groups

Diffraction group (with related chiral group)	Conditions as for	Additional conditions for $I(hkl) \equiv$	Additional conditions for $I(\bar{h}\bar{k}\bar{l}) \equiv$	Multiplicity of centrosymmetric general data
$\bar{1}$ (1)	–	–	–	2
$2/m$ (2)	$\bar{1}$ (1)	$I(\bar{h}k\bar{l})$	$I(h\bar{k}l)$	4
mmm (222)	$2/m$ (2)	$I(h\bar{k}\bar{l})$, $I(\bar{h}k\bar{l})$	$I(\bar{h}kl)$, $I(h\bar{k}l)$	8
$4/m$ (4)	$\bar{1}$ (1)	$I(\bar{h}\bar{k}l)$, $I(\bar{k}hl)$, $I(k\bar{h}l)$	$I(hk\bar{l})$, $I(k\bar{h}\bar{l})$, $I(\bar{k}h\bar{l})$	8
$4/mmm$ (422)	$4/m$ (4)	$I(\bar{h}k\bar{l})$, $I(h\bar{k}\bar{l})$, $I(\bar{k}h\bar{l})$, $I(k\bar{h}\bar{l})$	$I(h\bar{k}l)$, $I(\bar{h}kl)$, $I(k\bar{h}l)$, $I(\bar{k}hl)$	16
$\bar{3}$ (3)	$\bar{1}$ (1)	$I(kil)$, $I(ihl)$	$I(\bar{k}\bar{i}\bar{l})$, $I(\bar{i}\bar{h}\bar{l})$*	6
$\bar{3}m1$ (321)	$\bar{3}$ (3)	$I(\bar{k}\bar{h}\bar{l})$, $I(\bar{h}\bar{i}\bar{l})$, $I(\bar{i}\bar{k}\bar{l})$	$I(hil)$, $I(khl)$, $I(ikl)$	12
$\bar{3}1m$ (312)	$\bar{3}$ (3)	$I(kh\bar{l})$, $I(hi\bar{l})$, $I(ik\bar{l})$	$I(\bar{k}\bar{h}l)$, $I(\bar{h}\bar{i}l)$, $I(\bar{i}\bar{k}l)$	12
$6/m$ (6)	$\bar{3}$ (3)	$I(\bar{h}\bar{k}l)$, $I(\bar{k}\bar{i}l)$, $I(\bar{i}\bar{h}l)$	$I(hk\bar{l})$, $I(ih\bar{l})$, $I(ki\bar{l})$	12
$6/mmm$ (622)	$6/m$ (6)	$I(\bar{k}\bar{h}l)$, $I(\bar{h}\bar{i}l)$, $I(\bar{i}\bar{k}l)$; $I(kh\bar{l})$, $I(hi\bar{l})$, $I(ik\bar{l})$	$I(hil)$, $I(khl)$, $I(ikl)$; $I(\bar{k}\bar{h}l)$, $I(\bar{h}\bar{i}l)$, $I(\bar{i}\bar{k}l)$	24
$m\bar{3}$ (23)	mmm (222)	$I(klh)$, $I(\bar{k}lh)$, $I(k\bar{l}h)$, $I(\bar{k}\bar{l}h)$; $I(lhk)$, $I(\bar{l}hk)$, $I(l\bar{h}k)$, $I(\bar{l}\bar{h}k)$	$I(\bar{k}\bar{l}\bar{h})$, $I(k\bar{l}\bar{h})$, $I(\bar{k}l\bar{h})$, $I(kl\bar{h})$; $I(\bar{l}\bar{h}\bar{k})$, $I(l\bar{h}\bar{k})$, $I(\bar{l}h\bar{k})$, $I(lh\bar{k})$	24
$m\bar{3}m$ (432)	$m\bar{3}$ (23)	$I(lkh)$, $I(\bar{l}kh)$, $I(l\bar{k}h)$, $I(\bar{l}\bar{k}h)$; $I(khl)$, $I(\bar{k}hl)$, $I(k\bar{h}l)$, $I(\bar{k}\bar{h}l)$; $I(hlk)$, $I(\bar{h}lk)$, $I(h\bar{l}k)$, $I(\bar{h}\bar{l}k)$	$I(\bar{l}\bar{k}\bar{h})$, $I(l\bar{k}\bar{h})$, $I(\bar{l}k\bar{h})$, $I(lk\bar{h})$; $I(\bar{k}\bar{h}\bar{l})$, $I(k\bar{h}\bar{l})$, $I(\bar{k}h\bar{l})$, $I(kh\bar{l})$; $I(\bar{h}\bar{l}\bar{k})$, $I(h\bar{l}\bar{k})$, $I(\bar{h}l\bar{k})$, $I(hl\bar{k})$	48

*In trigonal and hexagonal crystals, $i = -h - k$.

is the subscript divided by the order of the axis, here always $\frac{1}{2}$. Fourfold axes can also involve translations of only $\frac{1}{4}$ of a unit cell. If this is in the sense of a right-handed screw, the axis is called 4_1; if left-handed 4_3. Similar operations involving a shift of one-third of a cell give 3_1, 3_2, 6_2 and 6_4 axes, and shifts of one-sixth give 6_1 and 6_5 axes. In practice, screw axes are much more common in crystallography than are pure rotation axes, as they help molecules to pack in a staggered arrangement relative to one another. Conventional representations of these axes in diagrams are shown in Table 3.6.

Because of the translational symmetry, reflections for crystal planes perpendicular to a screw axis will show 'absences' similar to those for Bravais lattices. These absences will only affect a single 'row' of data, usually only data with two indices zero (Table 3.7).

Table 3.5 Centred lattices and conditions for data

Bravais lattice	Symbol	Points equivalent to 0, 0, 0	Condition for data hkl	Fraction present
Primitive	P	none	none	1
A-centred	A	$0, \frac{1}{2}, \frac{1}{2}$	$k + l = 2n$	$\frac{1}{2}$
B-centred	B	$\frac{1}{2}, 0, \frac{1}{2}$	$h + l = 2n$	$\frac{1}{2}$
C-centred	C	$\frac{1}{2}, \frac{1}{2}, 0$	$h + k = 2n$	$\frac{1}{2}$
Body-centred	I	$\frac{1}{2}, \frac{1}{2}, \frac{1}{2}$	$h + k + l = 2n$	$\frac{1}{2}$
Face-centred	F	$\begin{cases} 0, \frac{1}{2}, \frac{1}{2} \\ \frac{1}{2}, 0, \frac{1}{2} \\ \frac{1}{2}, \frac{1}{2}, 0 \end{cases}$	$\begin{cases} h\,k\,l \\ \text{all odd or} \\ \text{all even} \end{cases}$	$\frac{1}{4}$
Rhombohedral	R	$\begin{cases} \frac{1}{3}, \frac{2}{3}, \frac{2}{3} \\ \frac{2}{3}, \frac{1}{3}, \frac{1}{3} \end{cases}$	$\begin{cases} -h + k + l \\ = 3n \end{cases}$	$\frac{1}{3}$

Table 3.6. Representations of rotation and screw axes

| 2 | 2_1 | 3 | 3_1 | 3_2 | 4 | 4_1 | 4_2 | 4_3 | 6 | 6_1 | 6_2 | 6_3 | 6_4 | 6_5 |

Table 3.7 Conditions for data resulting from screw axes

Axis parallel to	Row	$2_1, 4_2, 6_3$ axes	$3_1, 3_2, 6_2, 6_4$ axes	$4_1, 4_3$ axes	$6_1, 6_5$ axes
a	$h00$	$h = 2n$	$h = 3n$	$h = 4n$	$h = 6n$
b	$0k0$	$k = 2n$	$k = 3n$	$k = 4n$	$k = 6n$
c	$00l$	$l = 2n$	$l = 3n$	$l = 4n$	$l = 6n$

It is not possible to associate translation with any of the improper symmetry operations except reflection, where several possibilities arise, all of them being translations parallel to the plane of reflection. These are called glide planes and are usually of one of three types, depending on the direction of translation. In every case, the translation will be half way towards another lattice point, in most cases halfway along one or two of the crystallographic axes. Like screw axes, glide planes will give rise to absent data, and the effect will be much more general, as it will affect all reflections for crystal planes perpendicular to the glide plane, a group of data known as a 'zone'. The types of glides and the conditions are summarized in Table 3.8.

Table 3.8 Conditions for data arising from glide planes

Normal to glide	Direction of glide	Symbol	Zone affected	Condition
a	$y + \frac{1}{2}$	b	$0kl$	$k = 2n$
	$z + \frac{1}{2}$	c		$l = 2n$
	$y + \frac{1}{2}, z + \frac{1}{2}$	n		$k + l = 2n$
	$y + \frac{1}{4}, z + \frac{1}{4}$	d		$k + l = 4n$
b	$x + \frac{1}{2}$	a	$h0l$	$h = 2n$
	$z + \frac{1}{2}$	c		$l = 2n$
	$x + \frac{1}{2}, z + \frac{1}{2}$	n		$h + l = 2n$
	$x + \frac{1}{4}, z + \frac{1}{4}$	d		$h + l = 4n$
c	$x + \frac{1}{2}$	a	$hk0$	$h = 2n$
	$y + \frac{1}{2}$	b		$k = 2n$
	$x + \frac{1}{2}, y + \frac{1}{2}$	n		$h + k = 2n$
	$x + \frac{1}{4}, y + \frac{1}{4}$	d		$h + k = 4n$
$[110]$	$z + \frac{1}{2}$	c	hhl	$l = 2n$
	$x + \frac{1}{4}, y + \frac{1}{4}, z + \frac{1}{4}$	d		$2h + l = 4n$

3.6 Detection of symmetry elements from intensity statistics

The arguments so far will enable the lattice type, the diffraction symmetry, and those symmetry elements involving translation to be recognized. Symmetry elements which have no translational element and which are not required by the diffraction symmetry are more difficult to recognize. In particular, the presence or absence of an inversion centre is often strongly indicated by the intensity statistics. It must be held in mind, however, that statistical methods at best only indicate that something is *probably* true, and are based on the assumption of an essentially random arrangement of atoms. The presence of a few heavy atoms in a structure or of non-crystallographic molecular symmetry may sometimes make the statistical tests untrustworthy.

For all statistical tests, the intensity data, or the structure factors $|F(hkl)|$, are converted to normalized structure factors $|E(hkl)|$. This may be done approximately without any knowledge of the unit cell contents by using a curve of the mean value of $|F|^2$, $\langle |F|^2 \rangle_\theta$, for ranges of $\sin \theta$. The normalized structure factors are produced as $|E|^2(h\,k\,l) = |F|^2(h\,k\,l)/\langle |F|^2 \rangle_\theta$ and should have a mean value of $\langle |E|^2 \rangle = 1$ for all values of $\sin \theta$. The distribution of E-values about the mean in centrosymmetric structures will differ from that in non-centrosymmetric ones. This difference can be indicated in several ways.

In a structure in the triclinic space group $P1$, intensities will tend to a mean value, and the diffraction pattern will often look quite featureless. In $P\bar{1}$, however, the relationship of atoms in pairs will tend to give more very weak and very strong intensities. In terms of the probability $P(E)$ of a reflection having a particular value of $|E|$, the distributions

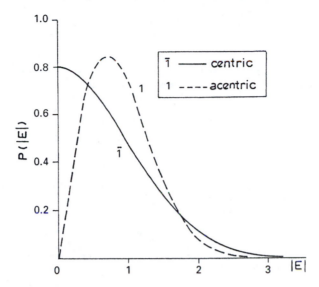

Fig. 3.2.   The distribution of intensities predicted for centrosymmetric and non-centrosymmetric structures.

may be given as:

$$\text{acentric } (P1) \qquad\qquad \text{centric } (P\bar{1})$$
$$P(E) = |E|\exp(-|E|^2) \qquad P(E) = \exp(-|E|^2/2)$$

The functions are shown graphically in Fig. 3.2.

In practice, the distributions may be used in various ways. The simplest is to calculate the value of $\langle |E|^2 - 1\rangle$ for all data. This will tend to 0.74 for the acentric and 0.97 for the centric distribution. Another approach is to calculate the proportion of data with $|E|^2$ below a certain value, the so-called $N(z)$ test. This gives a different range of values as follows:

% of $|E|^2<$	0.1	0.2	0.4	0.6	0.8	1.0
acentric	9.5	18.1	33.0	45.1	55.1	63.2
centric	24.8	34.5	47.4	56.1	62.9	68.3

The use of these statistics can be applied to other symmetry elements by noting that a twofold axis is equivalent to an inversion centre in projection, and so will give a centric distribution in the zone normal to it. Similarly, although less usefully, a mirror plane will give a centric distribution in a row normal to it. More significantly, the mirror plane will cause the reflections in its parallel zone to have twice the normal intensity, since atoms related by a plane normal to b will scatter X-rays like half the number of atoms with double the atomic number for all data in the $h0l$ zone. For all symmetry elements not involving translation, the intensity distributions and enhanced intensities are given in Table 3.9.

Often, the distinction between a centric and an acentric distribution is evident from the appearance of the diffraction pattern, as shown in Fig. 3.3.

Table 3.9 Distribution statistics and symmetry enhancements

Symmetry element	General data $(h\,k\,l)$		Perpendicular zone $(h\,k\,0)$		Parallel row $(0\,0\,l)$	
	Dist.	Enh.	Dist.	Enh.	Dist.	Enh.
1	A	1	A	1	A	
$2, 2_1$	A	1	C	1	A	2
$3, 3_1, 3_2$	A	1	A	1	A	3
$4, 4_1, 4_2, 4_3$	A	1	C	1	A	4
$6, 6_1, 6_2, 6_3, 6_4, 6_5$	A	1	C	1	A	6
$\bar{1}$	C	1	C	1	C	1
$\bar{2}\ (m, a, b, c, n, d)$	A	1	A	2	C	1
$\bar{3}$	C	1	C	1	C	3
$\bar{4}$	A	1	C	1	C	2
$\bar{6}$	A	1	A	2	C	3

A, acentric; C, centric—elements assumed parallel to c.

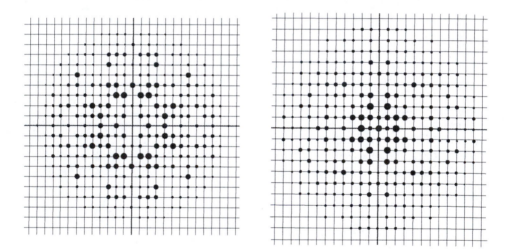

Fig. 3.3. Sections of the diffraction pattern for ammonium oxalate monohydrate ($P2_12_12$) showing (left) a centric distribution ($h\,k\,0$) and (right) an acentric distribution ($h\,k\,1$).

By these methods, it is possible in favourable circumstances to determine nearly all space groups uniquely. The ones remaining are the enantiomorphous pairs (e.g. $P3_1$ and $P3_2$) and the pairs $I222$ and $I2_12_12_1$, and $I23$ and $I2_13$.

3.7　Further notes on space group symbols

The form of the space group symbol as usually given is often an abbreviation, but may simply be described as follows. It consists of the Bravais lattice type followed by the point group symbol altered to indicate the translational symmetry elements present. The rules for doing this are:

triclinic: no translational symmetry—groups $P1$ and $P\bar{1}$.
monoclinic: symmetry along b given—2, m or $2/m$, altered to show 2_1 axes and a, c or n glide planes.
orthorhombic: symmetry along a, b and c axes. $2/m$ is abbreviated to m.
tetragonal, trigonal and hexagonal: symmetry along c is shown first; then, if there is any higher than $\bar{1}$, along a and b and finally along the ab-diagonal ([1 1 0]).
cubic: symmetry is given first along c, a and b, then along the body diagonals and finally, if there is any higher than $\bar{1}$, along the face diagonals.

Space groups are far from equally common. Molecular solids tend to concentrate in a very small number, and the 'top 6' make up more than 80% of all structures solved. Table 3.10 shows some of these statistics, which are useful in determining the likelihood of a particular space group.

Table 3.10 Frequency data concerning space groups from the Cambridge Structural Database

Rank	Space group	Number	Class	System	Totals 1999	(1986)	%	Special positions
1	$P2_1/c$	14	D_{2h}	mono	67012	(19561)	36.0	$\bar{1}$
2	$P\bar{1}$	2	C_i	tricl	37383	(8878)	17.6	$\bar{1}$
3*	$P2_12_12_1$	19	D_2	ortho	16901	(5875)	10.2	–
4	$C2/c$	15	C_{2h}	mono	13729	(3728)	7.0	$\bar{1}, 2$
5*	$P2_1$	4	C_2	mono	10632	(3472)	5.7	–
6	$Pbca$	61	D_{2h}	ortho	7179	(2267)	4.1	$\bar{1}$
7	$Pnma$	62	D_{2h}	ortho	2909	(918)	1.7	$\bar{1}, m$
8	$Pna2_1$	33	C_{2v}	ortho	2906	(963)	1.7	–
9	Cc	9	C_s	mono	1891	(530)	1.0	–
10*	$P1$	1	C_1	tricl	1846	(646)	1.3	–
11	$Pbcn$	60	D_{2h}	ortho	1686	(558)	1.0	$\bar{1}, 2$
12*	$C2$	5	C_2	mono	1541	(485)	0.9	2
13	$Pca2_1$	29	C_{2v}	ortho	1364	(425)	0.8	–
14	$P2_1/m$	11	C_{2h}	mono	1269	(385)	0.7	$\bar{1}, m$
15	$P2/c$	13	C_{2h}	mono	973	(260)	0.5	$\bar{1}, 2$
16	$C2/m$	12	C_{2h}	mono	972	(309)	0.6	$\bar{1}, 2, m, 2/m$
17	$R\bar{3}$	148	S_6	trig	937	(234)	0.4	$\bar{1}, 3, \bar{3}$
18*	$P2_12_12$	18	D_2	ortho	888	(293)	0.5	2
19*	$P4_12_12/P4_32_12$	92/96	D_4	tetrag	749	(267)	0.4	2
20	Pc	7	C_s	mono	691	(214)	0.4	–
21	$Pccn$	56	D_{2h}	ortho	664	(204)	0.4	$\bar{1}, 2$
22	$Fdd2$	43	C_{2v}	ortho	636	(193)	0.3	2
23	$I4_1/a$	88	C_{4h}	tetrag	615	(174)	0.3	$\bar{1}, 2, \bar{4}$

The total list in 1999 (1986) contained 186074 (54239) entries with space groups (no other space group has more than 500 entries, and most have 50 or less).
*Space groups possible for optically pure, chiral molecules.

Table 3.11   The 32 crystallographic point groups

Crystal system and Bravais lattices	Point groups Int.	Sch.	Space group nos.	Symmetry along x	y	z	Order[1]	E,P,C[2]
Triclinic	1	C_1	1	1	1	1	1	E,P
$P(=C, I, F)$	$\bar{1}$	C_i	2	$\bar{1}$	$\bar{1}$	$\bar{1}$	2	C
Monoclinic	2	C_2	3–5	1	2	1	2	E,P
P	m	C_s	6–9	1	m	1	2	P
$C(=I, F)$	$2/m$	C_{2h}	10–15	$\bar{1}$	$2/m$	$\bar{1}$	4	C
Orthorhombic	222	D_2	16–24	2	2	2	4	E
P, C, I, F	$mm2$	C_{2v}	25–46	$4m$	m	2	4	P
	mmm	D_{2h}	47–74	$2/m$	$2/m$	$2/m$	8	C
				z	x, y	xy		
Tetragonal	4	C_4	75–80	4	1	1	4	E,P
	$\bar{4}$	S_4	81–82	$\bar{4}$	1	1	4	–
$P(=C)$	$4/m$	C_{4h}	83–88	$4/m$	$\bar{1}$	$\bar{1}$	8	C
$I(=F)$	422	D_4	89–98	4	2	2	8	E
	$4mm$	C_{4v}	99–110	4	m	m	8	P
	$\bar{4}2m^*$	D_{2d}	111–122	$\bar{4}$	2	m	8	–
	$\bar{4}m2^*$			$\bar{4}$	m	2		
	$4/mmm$	D_{4h}	123–142	$4/m$	$2/m$	$2/m$	16	C
Trigonal	3	C_3	143–146	3	1	1	3	E,P
	$\bar{3}$	S_6	147–148	$\bar{3}$	$\bar{1}$	$\bar{1}$	6	C
P or R	321^*	D_3	149–155	3	2	1	6	E
	312^*			3	1	2		
	$3m1^*$	C_{3v}	156–161	3	m	1	6	P
	$31m^*$			3	1	m		
	$\bar{3}m1^*$	D_{3d}	162–167	$\bar{3}$	$2/m$	$\bar{1}$	12	C
	$\bar{3}1m^*$			$\bar{3}$	$\bar{1}$	$2/m$		
Hexagonal	6	C_6	168–173	6	1	1	6	E,P
	$\bar{6}$	C_{3h}	174	$\bar{6}$	1	1	6	–
P	$6/m$	C_{6h}	175–176	$6/m$	$\bar{1}$	$\bar{1}$	12	C
	622	D_6	177–182	6	2	2	12	E
	$6mm$	C_{6v}	183–186	6	m	m	12	P
	$\bar{6}2m^*$	D_{3d}	187–190	$\bar{6}$	2	m	12	–
	$\bar{6}m2^*$			$\bar{6}$	m	2	12	–
	$6/mmm$	D_{6h}	191–194	$6/m$	$2/m$	$2/m$	24	C
				x, y, z	xyz	xy, yz, zx		
Cubic	23	T	195–199	2	3	1	12	E
	$m\bar{3}$	T_h	200–206	$2/m$	$\bar{3}$	$\bar{1}$	24	C
P, I, F	432	O	207–214	4	3	2	24	E
	$\bar{4}3m$	T_d	215–220	$\bar{4}$	3	m	24	–
	$m\bar{3}m$	O_h	221–230	$4/m$	$\bar{3}$	$2/m$	48	C

[1] Multiplied by 2 for I or C lattices, by 3 for R and by 4 for F.

[2] Enantiomorphous, Polar or Centrosymmetric.

*In these point groups, different space groups occur because of the different possibilities for the arrangement of the symmetry elements.

3.8 Symmetry restrictions on atoms in special positions

A further direct consequence of the symmetry of a crystal concerns the refinement of some parameters of atoms in the structure. The parameters of an atom are restricted in accordance with the point group of the site on which it is located. There are three types of parameter to be considered: the site multiplicity, the positional parameters x, y and z, and the tensor describing the atomic displacement, U^{ij}.

The **multiplicity** of a site. An atom in a general position, unrestricted by symmetry, is said to have a multiplicity of 1, whatever the number of equivalent atoms in the unit cell (see Table 3.10). If an atom is in a special position, its multiplicity is the reciprocal of the order of the point group of the site. Thus, an atom on a site of mmm symmetry (order $= 8$) in any space group is said to have a multiplicity of $\frac{1}{8}$.

The **positional parameters**. These are unrestricted only for atoms in general positions. Otherwise, they are restricted as follows, where 0 is any fixed value; e.g. possibly $\frac{1}{2}$ or $\frac{1}{4}$:

(a) restricted to a plane (site symmetry m):

plane normal to:	$[1\,0\,0]$	$[0\,1\,0]$	$[0\,0\,1]$	$[0\,1\,1]$	$[1\,0\,1]$	$[1\,1\,0]$
atom restricted to:	$0, y, z$	$x, 0, z$	$x, y, 0$	$x, y, \bar{y}$	$x, y, \bar{x}$	$x, \bar{x}, z$

(b) restricted to a line (site symmetry 2, 3, 4, 6, $mm2$, $3m$, $4mm$, $6mm$):

axis parallel to:	$[1\,0\,0]$	$[0\,1\,0]$	$[0\,0\,1]$	$[0\,1\,1]$	$[1\,0\,1]$	$[1\,1\,0]$	$[1\,1\,1]$
atom restricted to:	$x, 0, 0$	$0, y, 0$	$0, 0, z$	$0, y, y$	$x, 0, x$	$x, x, 0$	x, x, x

(c) any other point group, the atom is fixed to a point.

The **atomic displacement parameters**. There are five basic possibilities for restrictions on the tensor that describes the displacement ellipsoid. These correspond to the restrictions on unit cell parameters in Table 3.3. (Slightly more complex forms apply to trigonal and hexagonal crystals.)

(a) shape and orientation unrestricted: (triclinic) point groups 1 and $\bar{1}$; 6 parameters.
(b) shape unrestricted, one of the principal axes of the ellipse parallel to a given direction, others normal to it: (monoclinic) point groups 2, m and $2/m$; 4 parameters

	U^{11}	U^{22}	U^{33}	U^{23}	U^{13}	U^{12}
$[1\,0\,0]$	A	B	C	D	0	0
$[0\,1\,0]$	A	B	C	0	D	0
$[0\,0\,1]$	A	B	C	0	0	D
$[0\,1\,1]$	A	B	B	C	D	D
$[1\,0\,1]$	A	B	A	C	D	C
$[1\,1\,0]$	A	A	B	C	C	D

(c) shape unrestricted, orientation fixed: (orthorhombic) point groups 222, *mm2* and *mmm*.

	U^{11}	U^{22}	U^{33}	U^{23}	U^{13}	U^{12}
$[1\,0\,0], [0\,1\,0], [0\,0\,1]$	A	B	C	0	0	0
$[1\,0\,0], [0\,1\,1], [0\,1\,\bar{1}]$	A	B	B	0	C	C
$[0\,1\,0], [1\,0\,1], [1\,0\,\bar{1}]$	A	B	A	C	0	C
$[0\,0\,1], [1\,1\,0], [1\,\bar{1}\,0]$	A	A	B	C	C	0

(d) shape restricted to one circular cross-section (trigonal, tetragonal or hexagonal)

	U^{11}	U^{22}	U^{33}	U^{23}	U^{13}	U^{12}
$[0\,0\,1]$	A	A	B	0	0	0
$[1\,1\,1]$	A	A	A	B	B	B

(e) shape restricted to a sphere—(cubic)—isotropic.

Exercises

3.1   Molecular symmetry and point groups. Examine the diagrams of the following molecules (Fig. 3.4), or better, make up models of them, and match each molecule with one of the point groups on the next page, which involve only the operations 2, *m* and $\bar{1}$:

(a) bromochlorofluoromethane (CHBrClF); (b) hypochlorous acid (HOCl); (c) ethene (C_2H_4); (d) *Z*-1,2-difluoroethene; (e) *E*-1,2-difluoroethene; (f) (1*R*,2*S*)1,2-dichloro-1,2-difluororethane (ClFHC−CClFH) (fully staggered); (g) (1*R*,2*R*)1,2-dichloro-1,2-difluororethane (ClFHC−CClFH) (any conformation); (h) (3*R*,4*R*,3′*R*,4′*R*)3,4,3′,4′-tetrachlorospirane ($C_9H_{12}Cl_4$).

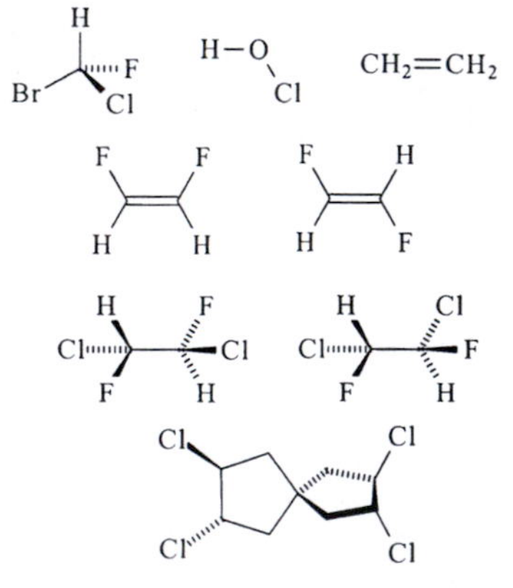

Fig. 3.4.   Molecules for Exercise 3.1.

3.2   Try to find the symmetry elements in the illustrations of crystal morphology (Fig. 3.1), and identify them with those in the point group table on the next page.

| Point group | | | | Twofold axis along | | | Mirror normal to | | |
Int.	Sch.	Identity	Centre	x	y	z	x	y	z
1	C_1	Y	n	n	n	n	n	n	n
$\bar{1}$	C_i	Y	Y	n	n	n	n	n	n
2	C_2	Y	n	n	n	Y	n	n	n
m	C_s	Y	n	n	n	n	n	n	Y
$2/m$	C_{2h}	Y	Y	n	n	Y	n	n	Y
222	D_2	Y	n	Y	Y	Y	n	n	n
$mm2$	C_{2v}	Y	n	n	n	Y	Y	Y	n
mmm	D_{2h}	Y	Y	Y	Y	Y	Y	Y	Y

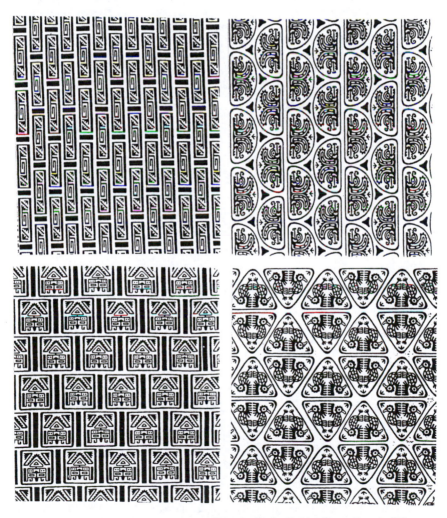

Fig. 3.5. Gift wrapping paper patterns (from the charity Traidcraft): see Exercise 3.5.

3.3 The following unit cell volumes and densities have been measured for the given compounds. Calculate Z for the crystal, and comment on how well (or badly) the '18 Å^3' rule works for each compound.
  (a) Methane (CH$_4$) [at 70 K]: $V = 215.8$ Å^3; $\rho = 0.492$ g cm^{-3}
  (b) Diamond (C): $V = 45.38$ Å^3; $\rho = 3.512$ g cm^{-3}
  (c) Glucose (C$_6$H$_{12}$O$_6$): $V = 764.1$ Å^3; $\rho = 1.564$ g cm^{-3}
  (d) Bis(dimethylgloxime)platinum(II) (C$_8$H$_{14}$N$_4$O$_4$Pt): $V = 1146$ Å^3; $\rho = 2.46$ g cm^{-3}.

3.4 A tetragonal crystal has $a(=b) = 10.48$ Å, $c = 6.05$ Å. Assuming that the Laue symmetry is $4/m$ and that data are required to a resolution of 0.8Å, estimate the number of unique data.

3.5 'Plane groups' are two-dimensional space groups, and the patterns used in wallpapers are good analogies for crystal structures. In two dimensions, there are five Bravais lattices (oblique, rectangular p, rectangular c, square, and hexagonal). Possible symmetry operations are two-, three-, four-, and sixfold rotation points, and lines of reflection, which may be either simple reflections or *glides* in which reflection is accompanied by a translation of $\frac{1}{2}$ along the line. These are represented in diagrams as solid lines (——) and broken lines (– – –) respectively. In each of the wrapping paper patterns in Fig. 3.5, find a unit cell and identify as many symmetry elements as you can. Finally, using the simple diagrams for the oblique and rectangular patterns given in Fig. 3.6 or those in the *International Tables for Crystallography*, pair up each pattern with its appropriate plane group, and identify the asymmetric unit of the pattern.

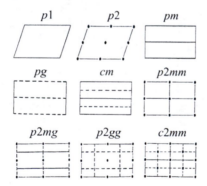

Fig. 3.6. Selected plane groups (two-dimensional space groups), showing symmetry elements.

3.6 Deduce as much as you can about the space group of the compounds for which the following data were obtained.
  (a) Orthorhombic P. Conditions: $h0l$: $h = 2n$; $0kl$: $k + l = 2n$. General data give acentric distribution, $hk0$ give centric, $h0l$ and $0kl$ give twice the normal average intensity.
  (b) Tetragonal P. Conditions: $h00$: $h = 2n$ ($0\,k\,0$: $k = 2n$); $00l$: $l = 4n$. $I(hkl) = I(khl)$. General data acentric, $hk0$, $h0l$, $0kl$ and hhl zones centric. Would you expect any symmetry enhancement of zones or rows?
  (c) Monoclinic C. Conditions $h0l$: $l = 2n$; zone $h0l$ gives twice the normal average intensity, but other intensity statistics are ambiguous. Compound is thought to be a nickel complex of the form [NiL^{2+}]Cl$_2^-$, where L is 1,4,7,10-tetraazacyclododecane. There are four molecules per unit cell.
  (d) Apparently triclinic P. Cell dimensions: $a = 3.952$, $b = 6.772$, $c = 9.993$ Å, $\alpha = 98.06$, $\beta = 89.96$, $\gamma = 106.96°$. Four molecules per unit cell. Inspection of intensity

pattern shows that the following pairs of reflections are strong with equal intensity and $\sin\theta$ values: 123 and $\bar{1}33$; 240 and $\bar{2}60$; 440 and $\bar{4}00$. Curiously, for a triclinic cell, all data are absent for the zone $0kl$ with l odd. A plot to scale of the lattice projected along [001] is shown below (Fig. 3.7).

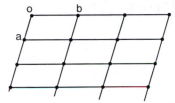

Fig. 3.7. Projection of the crystal lattice for Exercise 3.6(d).

4

Background theory for data collection

4.1 Introduction

Although many diffractometers are operated, and crystal structures determined, by people with little knowledge of the theory underlying the subject, success is more likely and problems will be better avoided by some understanding of the fundamental properties of the crystalline solid state, the nature of diffraction, and the relationships between a crystal structure and its diffraction pattern. Some of these topics are dealt with under the heading of symmetry. Here we consider particularly the geometry of diffraction.

An X-ray diffraction pattern has a particular geometry: it consists of individual scattered X-ray beams, each in a certain specific direction. These may be measured one at a time by a four-circle diffractometer, by positioning the detector appropriately for each diffracted beam generated by a certain crystal orientation relative to the X-ray source, or they may be recorded on a photographic film or an electronic area detector as a pattern of spots, the positions of which are not random. The geometry of the diffraction pattern is related to the lattice and unit cell geometry of the crystal structure, and so can tell us the repeat distances between molecules.

The pattern also has symmetry, which is closely related to the symmetry of the unit cell of the crystal structure, i.e. to the crystal system and space group.

Apart from the symmetry, there is no apparent relationship among the intensities of the individual diffracted beams, which vary widely; some are very intense, while others are too weak to be detected above the general background level (their positions are deduced from the geometrical regularity of the diffraction pattern). These intensities hold all the available information about the positions of atoms in the unit cell of the crystal structure. Thus, determination of the full molecular structure involves the measurement of all the many individual intensities, and these intensities can only be measured when the directions of the diffracted beams (the diffraction pattern geometry) have first been established.

4.2 The geometry of X-ray diffraction

Consider diffraction by a single row of regularly spaced points (one-dimensional diffraction; see Fig. 4.1). In any particular direction, the radiation scattered by the row of points will have zero intensity by destructive interference of the individual scattered rays unless they are all in phase. Since, except in the straight-through direction, individual rays have different path lengths, these path differences must be equal to whole numbers of

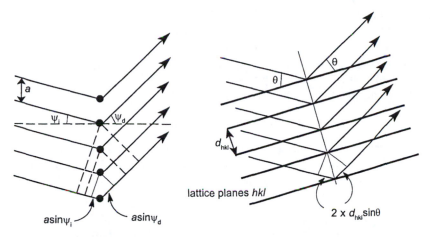

Fig. 4.1.   The Laue and Bragg constructions for X-ray diffraction.

wavelengths to keep the rays in phase. So, for rays scattered by two adjacent points in the row

$$\text{path difference} = a \sin \psi_i + a \sin \psi_d = h\lambda \qquad (4.1)$$

where ψ_i and ψ_d are the angles of the incident and diffracted beams as shown, λ is the wavelength, a is the one-dimensional lattice spacing, and h is an integer (positive, zero or negative). For a given value of ψ_i (a fixed incident beam), each value of h corresponds to an observed diffraction maximum and the equation can be used to calculate the permitted values of ψ_d, the directions in which intensity is observed. The result is a set of bright fringes.

For a significant diffraction effect, the spacing a must be comparable to the wavelength λ. This is why X-rays are used for diffraction by crystals.

For diffraction by a three-dimensional lattice there are three such equations and all have to be satisfied simultaneously. The first equation contains the lattice a spacing, angles relative to this a axis of the unit cell, and an integer h. The other two equations, correspondingly, contain the unit cell axes b and c, and integers k and l respectively.

Thus each allowed diffracted beam (each spot seen in an X-ray diffraction pattern) can be labelled by three integers, or indices, hkl, which uniquely specify it if the unit cell geometry is known.

These three equations for diffraction geometry, the Laue conditions, are cumbersome to use in this form. An alternative but equivalent description was derived by W. L. Bragg soon after the experimental demonstration that X-rays could be diffracted by crystals, and is expressed in the single Bragg equation, which is universally used as the basis for X-ray diffraction geometry. Bragg showed that every diffracted beam that can be produced by an appropriate orientation of a crystal in an X-ray beam can be regarded geometrically as if it were a reflection from sets of parallel planes passing through lattice points, analogous to the reflection of light by a mirror, in that the angles of incidence and reflection must be equal and that the incoming and outgoing beams and the normal to the reflecting planes must themselves all lie in one plane. The reflection by adjacent planes in the set gives interference effects equivalent to those of the Laue equations; to define a plane we need three integers to specify its orientation with respect to the three unit cell edges, and these are the indices hkl; and the spacing between successive planes is determined by the lattice geometry, so is a function of the unit cell parameters.

For rays reflected by two adjacent parallel planes

$$\text{path difference} = 2d_{hkl} \sin \theta = n\lambda. \tag{4.2}$$

In practice, the value of n can always be set to one by considering planes with smaller spacing ($n = 2$ for planes hkl is equivalent to $n = 1$ for planes $2h, 2k, 2l$, which have exactly half the spacing) and it is in the form

$$\lambda = 2d_{hkl} \sin \theta \tag{4.3}$$

that the Bragg equation is always used. It allows each observed diffracted beam (commonly known as a 'reflection') to be uniquely labelled with its three indices and for its net scattering angle (2θ from the direct beam direction) to be calculated from the unit cell geometry, of which each d_{hkl} spacing is a function.

Rearrangement of the Bragg equation gives

$$\sin \theta = \frac{\lambda}{2} \cdot \frac{1}{d_{hkl}}. \tag{4.4}$$

The distance of each spot from the centre of an X-ray diffraction pattern is proportional to $\sin \theta$ and hence to $1/d_{hkl}$ for some set of lattice planes. This demonstrates mathematically the reciprocal (inverse) nature of the geometrical relationship between a crystal lattice and its diffraction pattern.

4.3 The reciprocal lattice

The reciprocal relationship seen in the Bragg equation, together with the associated geometrical conditions, leads to a mathematical construction called the reciprocal lattice, which provides an elegant and convenient basis for calculations involving diffraction geometry. The direct lattice is the lattice of the crystal structure, defined by three vectors $\mathbf{a}$, $\mathbf{b}$, $\mathbf{c}$ (encompassing the six unit cell parameters, since vectors have both magnitude and direction). The reciprocal lattice is defined by three vectors $\mathbf{a}^*$, $\mathbf{b}^*$, $\mathbf{c}^*$, and is related to the direct lattice by the following equations:

$$\mathbf{a}^* = \frac{\mathbf{b} \times \mathbf{c}}{V} \qquad \mathbf{b}^* = \frac{\mathbf{c} \times \mathbf{a}}{V} \qquad \mathbf{c}^* = \frac{\mathbf{a} \times \mathbf{b}}{V}$$

$$V = \mathbf{a} \cdot (\mathbf{b} \times \mathbf{c}) = \mathbf{b} \cdot (\mathbf{c} \times \mathbf{a}) = \mathbf{c} \cdot (\mathbf{a} \times \mathbf{b})$$

$$\text{Hence: } \mathbf{a} \cdot \mathbf{a}^* = \mathbf{b} \cdot \mathbf{b}^* = \mathbf{c} \cdot \mathbf{c}^* = 1$$

$$\text{and } \mathbf{a} \cdot \mathbf{b}^* = \mathbf{a} \cdot \mathbf{c}^* = \mathbf{b} \cdot \mathbf{a}^* = \mathbf{b} \cdot \mathbf{c}^* = \mathbf{c} \cdot \mathbf{a}^* = \mathbf{c} \cdot \mathbf{b}^* = 0. \tag{4.5}$$

Each of these equations remains valid if all starred quantities become unstarred and all unstarred quantities become starred.

The reciprocal lattice is a convenient representation of the geometry of the diffraction pattern, with each reciprocal lattice point representing one Bragg reflection. The three coordinates of a reciprocal lattice point, counting from the centre, along the a^*, b^* and c^* axes in turn, are just the reflection indices h, k, l. The distance of a reciprocal lattice point from the origin is the length of the vector $(h\mathbf{a}^* + k\mathbf{b}^* + l\mathbf{c}^*) = 1/d_{hkl} = (2 \sin\theta)/\lambda$ from the Bragg equation. In fact, one particular type of X-ray photograph, produced by a precession camera, is a directly scaled undistorted picture of a section of the reciprocal lattice; the regular spacings between reflections are directly proportional to reciprocal cell axis lengths, and the angles between these rows are reciprocal cell angles.

Reciprocal cell parameters are used in diffractometer control programs to manipulate the diffraction geometry, deriving cell and crystal orientation parameters from selected observed reflections, and then using these to predict the positions of all reflections for intensity measurements.

4.4 Unit cell and orientation matrix on a diffractometer

Two sets of axes are of particular importance in diffractometry (both conventional four-circle machines and those with area detectors). The first is the crystal-fixed set of reciprocal lattice axes, a^*, b^* and c^*. In this axial system, the coordinates of a Bragg reflection are the Miller indices hkl (denoted by the vector $\mathbf{h}$). The second axis set is an orthogonal one fixed to the crystal mounting: the z axis is coincident with the diffractometer ϕ axis; x and y are defined arbitrarily to complete a right-handed set and the definitions are not the same for all machines. The relationship between the coordinates of any point referred to these two sets of axes is given by

$$\mathbf{x} = \mathbf{A}\,\mathbf{h} \tag{4.6}$$

where the vector $\mathbf{x}$ denotes the three coordinates xyz.

The orientation matrix $\mathbf{A}$ lies at the centre of the whole data collection process. Once it has been determined, the correct diffractometer angles and detector settings can be calculated for any reflection. The nine elements of the matrix $\mathbf{A}$ are the components of the reciprocal cell axes on each of the $\mathbf{x}$ axis set.

$$\mathbf{A} = \begin{pmatrix} a_x^* & b_x^* & c_x^* \\ a_y^* & b_y^* & c_y^* \\ a_z^* & b_z^* & c_z^* \end{pmatrix}. \tag{4.7}$$

These nine elements contain information about the unit cell (requiring six parameters) and orientation of the crystal (three parameters). The cell parameters are easily extracted:

$$(\mathbf{A}'\mathbf{A})^{-1} = \begin{pmatrix} \mathbf{a}\cdot\mathbf{a} & \mathbf{a}\cdot\mathbf{b} & \mathbf{a}\cdot\mathbf{c} \\ \mathbf{b}\cdot\mathbf{a} & \mathbf{b}\cdot\mathbf{b} & \mathbf{b}\cdot\mathbf{c} \\ \mathbf{c}\cdot\mathbf{a} & \mathbf{c}\cdot\mathbf{b} & \mathbf{c}\cdot\mathbf{c} \end{pmatrix} \tag{4.8}$$

where $\mathbf{A}'$ is the transpose of the $\mathbf{A}$ matrix.

The calculation of the four-circle diffractometer angles from the xyz coordinates of a reflection is trigonometric, and the details depend on the sign conventions for each particular machine. For the usual bisecting geometry ($\omega = \theta$), typical equations are:

$$\varphi = \tan^{-1}\left(\frac{-x}{y}\right)$$

$$\chi = \tan^{-1}\left(\frac{z}{\sqrt{x^2 + y^2}}\right) \tag{4.9}$$

$$\omega = \theta = \sin^{-1}\left(\frac{\lambda\sqrt{x^2 + y^2 + z^2}}{2}\right).$$

Each reflection can, however, be brought into the Bragg position in any number of ways (providing there are no mechanical obstructions). This can be thought of as an arbitrary rotation about the diffraction vector (which bisects 2θ) by an angle ψ (known as the azimuthal angle and defined to be zero in the bisecting position). More general equations for the diffractometer angles, of which the above are special cases, are then:

$$r = \sqrt{x^2 + y^2 + z^2}$$

$$\varphi = \tan^{-1}\left(\frac{-x}{y}\right) + \tan^{-1}\left(\frac{-r\sin\psi}{|z|\cos\psi}\right)$$

$$\chi = \tan^{-1}\left[\frac{z}{y\cos\varphi - x\sin\varphi}\right] \tag{4.10}$$

$$\omega = \theta + \tan^{-1}\left[\frac{x\cos\varphi + y\sin\varphi}{z/\sin\chi}\right]$$

$$\theta = \sin^{-1}\left(\frac{\lambda r}{2}\right).$$

Diffractometers operating on kappa geometry rather than with a Eulerian cradle have equivalent equations for the calculation of their angles; in some cases, they are expressed also in standard Eulerian geometry on these instruments.

These interconversions may all be carried out in reverse: the indices hkl can be derived from xyz coordinates by

$$\mathbf{h} = \mathbf{A}^{-1}\mathbf{x} \tag{4.11}$$

and xyz from the angles which centre a reflection in the counter by:

$$x = r(\cos\phi\sin\omega' - \sin\phi\cos\chi\cos\omega')$$

$$y = r(\sin\phi\sin\omega' + \cos\phi\cos\chi\cos\omega')$$

$$z = r(\sin\chi\cos\omega') \tag{4.12}$$

$$\text{where } \omega' = \omega - \theta$$

$$\text{and } r = 2(\sin\theta)/\lambda.$$

The process of setting up a crystal for data collection involves finding reflections, assigning indices to these and determining the unit cell and orientation, refining and checking the unit cell, and determining the Laue symmetry.

Similar procedures apply for area detector systems, although the details of the calculations are obviously different and involve the positions of reflections on the face of the detector (and various detector correction factors) as well as angles for the various motors.

On all modern diffractometers there is a considerable degree of automation in these procedures, and some machines can carry out the whole process of setting up and data collection without human intervention once a suitable crystal has been mounted and optically centred. Even in such cases (or perhaps especially in such cases!) an understanding of what is happening is important, so that problems can be avoided or solved.

4.5   Obtaining a matrix and cell from initially found reflections

Of the two basic methods used for determination of orientation matrix and unit cell from a set of unindexed reflections, one operates in reciprocal space [1], the other in real space [2]. They both aim to find the simplest cell and matrix which allows integral indices to be assigned to all the reflections. The reciprocal space method is more easily understood. The observed vectors $\mathbf{x}$ must correspond to reciprocal lattice points. To the list of vectors obtained from the initial search are added all the vectors $\mathbf{x}_i \pm \mathbf{x}_j$, which also represent reciprocal lattice points. From the augmented list, the three shortest non-coplanar vectors are selected as the reciprocal cell axes; the nine xyz coordinates are the nine elements of $\mathbf{A}$. Now it is possible to generate the indices of all the reflections. If any indices are simple fractions, one or more of the reciprocal axes is too long and the necessary correction is made. Other indices not close to integers may be caused by twinning, by the presence of more than one single crystal, or by spurious reflections, and these need to be investigated before proceeding further. The unit cell obtained by this method is necessarily primitive, and may not immediately disclose the full metric symmetry (the symmetry of the lattice). After least-squares refinement on the basis of all the genuine reflections (see below), the conventional cell and metric symmetry can be determined, usually by an automatic procedure.

The real-space method (known as 'auto-indexing') has been described in algebraic terms by Sparks [3,4]. An alternative explanation [2] is instructive. If we take the three shortest non-coplanar $\mathbf{x}$ vectors, and arbitrarily assign to them indices 100, 010 and 001, we can generate an orientation matrix and unit cell from their indices and angles. Although this cell $(\mathbf{a}', \mathbf{b}', \mathbf{c}')$ is not usually the correct one, it must be a sub-cell: all vectors in the true crystal lattice are also vectors in the lattice described by the preliminary sub-cell, though the converse is not true. The auto-indexing method generates vectors $\mathbf{t} = u\mathbf{a}' + v\mathbf{b}' + w\mathbf{c}'$ (u, v, w integral) up to a specified maximum length and tests each one to see whether is could be a true lattice vector, for which $\mathbf{t} \cdot \mathbf{x}$ must be an integer within a suitable small tolerance for every reflection in the list. The correct unit cell axes can then be chosen by hand or automatically from the generated list of lattice vectors. Once again, a failure to index all the reflections by this method could be caused by various faults in the crystal or by spurious reflections.

A further enhancement of the auto-indexing method, particularly useful with area detectors (which provide a much larger set of initial reflections) and essential for very large unit cells, involves using 'difference vectors' instead of the set of reciprocal space vectors **x** corresponding to the reflections themselves. From the list of **x** vectors, all differences $\mathbf{x}_i - \mathbf{x}_j$ are generated. For a substantial list of reflections, many of these difference vectors will occur repeatedly, and such replicate vectors are averaged, thus improving their accuracy.

There are two problems in particular which may arise in the procedure for cell determination. First, if the reflections all belong to a special subset, an incorrect cell is likely to be found: either too small, or with a wrong cell centring. For example, if all reflections have even h, the proposed **a** axis will be half the correct length; if all reflections have $h + k + l$ even, an I-centred cell will be indicated. This is most likely to occur with heavy atoms in special positions, causing certain classes of reflections to be systematically weak, especially if only the very strongest reflections are chosen from a photograph. The danger is also increased if too few reflections are used in the calculations. This is much less likely to happen with area detector systems, which generally use far more reflections in the initial cell determination (typically over a hundred rather than 10–25).

The other potential problem is caused by the presence of spurious reflections (i.e. not belonging to the correct reciprocal lattice) in the list. These may be a result of instrument faults, multiple crystals or twinning, and can lead either to too large a cell or to no cell at all, because the false reflections cannot be made to fit. Allowance for some 'anomalous' reflections in the cell determination routine may help find an otherwise intractable cell, but such a provision should be used with extreme care, because it can produce a sub-cell, rejecting genuine reflections in order to do so [2]. 'Anomalous' reflections should be carefully investigated before they are rejected as spurious.

A procedure has been described which generates reliable cell and matrix information even from a very obstinate list of reflections, circumventing the problems of spurious or inaccurate reflections, twinning and other effects [5]. The method involves considerable calculation, but is generally successful where others fail. It works in both real and reciprocal space together, exploiting the relationship between the two. Every combination of three reflections in the list is taken, in turn, to define a plane in reciprocal space, perpendicular to which lies a potential real space lattice vector; the length of this vector is deduced by projecting all the reflections on to the vector direction and finding the reciprocal space repeat from the pattern produced. Non-fitting reflections can be recognized easily.

The matrix and cell thus obtained, by any of these methods, must be refined to give the best fit to all the available reflections and any poorly fitting reflections investigated. The next step is to examine the symmetry of the diffraction pattern.

4.6   Symmetry aspects of the diffraction pattern

The crystallographic literature contains many examples of unit cells and space groups initially reported incorrectly and subsequently revised. In many cases, these are probably a consequence of blind faith in automatic procedures for cell determination on a diffractometer.

It is important to distinguish the different meanings of 'symmetry' in reference to crystal structures: the complete collection of all symmetry elements of the structure is the space group; the symmetry of the diffraction pattern (giving the equivalence of different reflections) is the Laue group; and the symmetry displayed by the geometry of the crystal lattice (and the reciprocal lattice, without regard to the intensities of individual reflections) is the metric symmetry. The metric symmetry can never be lower than the Laue symmetry, but it can be higher (e.g. an orthorhombic cell with $a = b$ is metrically tetragonal at least). The most common mistake of this kind is the failure to recognize the presence of symmetry elements in the cell determination procedure. Thus, centred monoclinic cells may be reported as only triclinic, rhombohedral as monoclinic, etc. This is probably due in most cases to an automatic procedure which fails, because of wrong decisions concerning the 'equality' of numbers subject to experimental error and computational rounding effects. This, in turn, is likely to be caused by an over-optimistic estimate of the precision of the initially determined unit cell. It is probably a good idea to test the cell symmetry with the available software once more, when final refined parameters have been obtained. An independent check with a program employing a different algorithm is also recommended for complete assurance; a particularly elegant method is described by Le Page [6]. This, and other routines for checking symmetry (and many more useful calculations) are available in the general purpose crystallographic package PLATON [7], which is available also in a Windows version [8].

Because the metric symmetry may actually be higher than the Laue symmetry, the latter should be checked by comparing the intensities of supposedly equivalent reflections. Unfortunately, however, severe absorption can cause the intensities of equivalent reflections to differ enormously. In cases of uncertainty, it is best to assume a lower symmetry for data collection. If the higher symmetry is subsequently confirmed, the equivalent reflections can be merged together to give a more reliable unique data set.

References

[1] J. Hornstra and H. Vossers, *Philips Tech. Rundschau*, 1974, **33**, 65–78.

[2] W. Clegg, *J. Appl. Cryst.*, 1984, **17**, 334–336.

[3] R. A. Sparks, *Crystallographic Computing Techniques*, ed. F. R. Ahmed, Munskgaard, Copenhagen, 1976, pp. 452–467.

[4] R. A. Sparks, *Computational Crystallography*, ed. D. Sayre, Clarendon Press, Oxford, 1982, pp. 1–18.

[5] A. J. M. Duisenberg, *J. Appl. Cryst.*, 1992, **25**, 92–96.

[6] Y. Le Page, *J. Appl. Cryst.*, 1982, **15**, 255–259.

[7] http://www.crystal.chem.uu.nl/software/PLATON.html (A. L. Spek, University of Utrecht, The Netherlands).

[8] http://www.chem.gla.ac.uk/~louis/platon/index.html (L. J. Farrugia, University of Glasgow, UK).

Exercises

4.1   What are the relationships between direct and reciprocal lattice parameters in (a) the orthorhombic; (b) the monoclinic system?

4.2   From the Bragg equation, what is the minimum d-spacing that can be observed with
       (a) Mo Kα; (b) Cu Kα radiation?

4.3   What Bragg angle corresponds to a resolution (d spacing) of 0.84 Å for each of these
       radiations?

4.4   Take the general equations for 2θ, ω, χ, ϕ and simplify them for the special cases (a) $\psi = 0$;
       (b) $\psi = 90°$.

4.5   From the orientation matrix

$$\mathbf{A} = \begin{pmatrix} 0 & 0.250 & 0 \\ 0.125 & 0 & 0 \\ 0 & 0 & -0.100 \end{pmatrix}$$

       calculate the unit cell parameters. About which axis is the crystal mounted? Is this desirable?
       Use this matrix also for the remaining exercises.

4.6   Calculate the bisecting position angles for the reflections 200, 020, 002, $42\bar{5}$; assume
       $\lambda = 1$ Å.

4.7   Calculate the angles for 002 and $42\bar{5}$ with $\psi = 90°$. Are these positions likely to be
       accessible? What about the corresponding positions for reflections 200 and 020?

4.8   Calculate the indices of the reflection observed at $2\theta = 17.08$, $\omega = 8.54$, $\chi = -19.69$,
       $\phi = -63.43°$.

4.9   Suppose $2\theta_0$ is incorrect so that 2θ is observed to be 0.05° too low for all reflections. Use
       the Bragg equation to calculate the c axis length from the observed, incorrect 2θ value for
       the reflection 002 (use $\lambda = 1$ Å).

5

Data collection using four-circle diffractometers

5.1 Introduction

Unlike the other stages of structure determination, the collection of diffraction data occurs in real time, requiring the continuous and exclusive use of valuable equipment. While a refinement which has gone wrong can usually be repeated without causing significant delay, an abandoned or terminally flawed data collection has wasted instrument time irrevocably. The aim in collecting a dataset should always be to obtain the best possible quality of data from the available sample. This requires not only preparing or demanding the best possible crystal, a topic which has already been covered, but also choosing experimental conditions and parameters which are appropriate. This chapter covers practical data collection in general, with further specific reference to the use of traditional four-circle diffractometers; differences in approach with area detectors, which are increasingly being used instead, are considered in the next chapter.

5.2 Experimental conditions

5.2.1 Radiation

The crystallographer has some choice over the conditions under which data are collected and one of these is the wavelength of the radiation used, the most common choice being between copper ($\lambda = 1.54184$ Å) and molybdenum (0.71073 Å). Copper X-ray tubes produce a higher flux of incident photons (for the same power settings) and these are diffracted more efficiently than molybdenum radiation; copper radiation is therefore particularly useful for small or otherwise weakly diffracting crystals. For crystals with long unit cell dimensions, reflections are further apart when the longer wavelength copper radiation is used and this can minimize reflection overlap. (With small unit cells, more driving between successive reflections is necessary as reflections are so far apart in reciprocal space that data collection becomes very inefficient.) If you want to determine absolute configurations and your crystals contain only elements lighter than silicon, copper radiation is usually essential. On the other hand, absorption effects are generally less serious with molybdenum radiation and this can be crucial if elements of high atomic number are present. Molybdenum radiation allows collection of data to higher resolution and is likely to cause fewer restrictions if low-temperature or other attachments are required. Changing radiation requires some effort and skill and you lose data collection time. The ideal situation is to have two diffractometers, one equipped with each radiation, and a supply of suitable crystals so that both may be fully used. The following illustrative examples may be helpful:

Well-diffracting organic compound containing iodine	Mo to minimize absorption
Poorly diffracting organic (CHNO) compound	Cu to maximize diffracted intensity
Organic compound (CHNO) with $b > 30$ Å	Cu to minimize overlap
Absolute configuration on $C_{19}H_{29}N_3O_7$	*Must* use Cu
Most metal complexes	Mo to minimize absorption
High-resolution, low-temperature studies	Mo
Poorly diffracting platinum complex	A 'no-win' situation?

5.2.2   Temperature

If a reliable low-temperature system is available, it is almost always worthwhile considering data collection at low temperature (most devices will even produce slightly elevated temperatures if this is required for a special experiment). The benefits of low-temperature data collection can only be realized if the equipment is well aligned and correctly set up: for example, icing can lead to the crystal moving or even being lost. Reducing the temperature of the crystal can have many advantages and is essential for crystals mounted using protective oil films, compounds melting below about 50°C and those which are thermolabile. Reactive compounds may be stabilized long enough to allow data collection. There are general advantages: at lower temperatures thermal motion is reduced and the intensities of reflections at higher Bragg angles thereby enhanced, allowing the collection of better diffraction data at higher resolution. The reduction in thermal motion also minimizes librational effects which give artificially shortened bond lengths and other systematic errors. One example of this is the relative ease with which disorder can be modelled: in particular, popular pseudo-spherical anions such as BF_4^-, PF_6^-, ClO_4^- and SO_4^{2-} are usually badly disordered at room temperature but either they are ordered or their disorder is much easier to model at low temperature. The actual temperature chosen is usually a compromise between the desire for the lowest temperature and the increased risk of icing as the temperature is reduced. For routine work on molecular crystals temperatures in the range 120–200 K are typical. Phase changes appear to be a relatively rare problem, but cooling can have adverse effects on poor quality crystals, showing up in the splitting of reflections, in poor orientation matrices and in larger uncertainties on cell parameters. Sometimes the splitting can be annealed out by increasing the temperature, for example from 150 to 220 K. Even in these apparently unfavourable cases a low-temperature determination is often better than one at ambient temperature.

5.2.3   Other conditions

Considerations of crystal size, methods of mounting, choice of goniometer head and optical centring have been dealt with previously but are worth stressing again, as they can seriously affect the outcome of the experiment. The collimator selected must allow the entire crystal to be immersed in the X-ray beam, but its diameter should not be excessive, as this will contribute to scattering by the air, resulting in increased background levels. This is more serious with copper than with molybdenum radiation.

5.3 Getting started

5.3.1 Reflection searching

It is rare these days for a crystal to be photographed before being put on the diffractometer, since a complete set of photographs can take longer than the data collection. You will normally start off with no information as to crystal system or unit cell dimensions and care is required to ensure that these are determined correctly. There are two basic methods available for locating the reflections required to determine the initial orientation matrix and unit cell on a four-circle diffractometer, the basic construction of which is shown schematically in Fig. 5.1.

The first of these is essentially a blind searching procedure: typically, the diffractometer control software drives the three circles 2θ, ω and χ (or 2θ, ω and κ) to randomly generated values, then moves ϕ through a range, usually of $180°$ or $360°$. If a reflection is detected it is either centred immediately or its position stored so that it and any others detected can be centred at the end of the ϕ scan. There are two variants on this method, one where a systematic stepwise search is undertaken, and one where each ϕ scan is performed at new, random angular positions. The latter is preferable as it gives a wider angular range of centred reflections and may result in fewer problems during

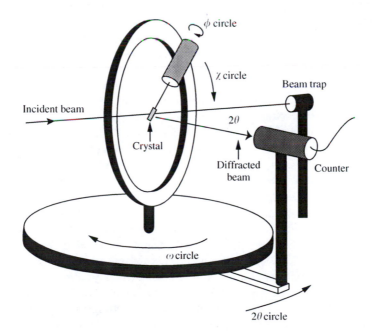

Fig. 5.1. Schematic diagram of the main features of a Eulerian cradle four-circle diffractometer.

cell determination. If only the former is implemented on your diffractometer, you may have to intervene to ensure that your list of centred reflections represents a wide range of values in χ (or κ). The search procedure is controlled by various parameters, including those specifying angular ranges, ϕ stepping rate, and the required peak to background ratio. As mentioned above, a wide range of ϕ and χ (or κ) values is desirable. The 2θ range (e.g. 10–20° for molybdenum, 20–40° for copper radiation), ϕ stepping rate (1–$2°$ s^{-1}), and required peak to background ratio (3.0) may all have to be reduced for weakly diffracting crystals. In these circumstances the blind search procedure becomes less efficient and photographic methods should be considered. Conversely, if you have a strongly diffracting crystal where short cell axes are expected you could increase the 2θ range.

The alternative method is photographic and involves making a rotation or oscillation photograph using a stationary film. Polaroid film requires shorter exposure and (especially) processing times and is more convenient than normal X-ray film. When the film is ready, the positions of any reflections are measured and input into the diffractometer control program, which can calculate all the setting angles apart from ϕ. A search on ϕ will then quickly locate each reflection. The accurate crystal-to-film distance is required for the calculation of angles and must therefore be known directly or by measurement of fiducial marks made on the film by very short exposures using an attenuated direct beam.

There are advantages to both methods. The random or systematic search can be left to run without intervention and within an hour may have generated enough reflections for indexing. The photographic method is more labour-intensive but faster, especially with poor crystals, small unit cells or copper radiation. Non-diffracting samples and multiple crystals can be identified as soon as the film is ready (within 10 minutes with a Polaroid cassette). Crucially, an estimate of the quality of a crystal (and a decision about whether to proceed with it) can be made at a very early stage. It may be possible to use a combination of these two methods to assemble a list of centred reflections for indexing.

Once reflections have been located, by whatever means, they must be centred. This involves adjusting the settings of the circles so that the diffracted beam at its maximum intensity passes into the centre of the detector. During this procedure ϕ is normally kept fixed as an ω scan moves the crystal through the Bragg position, giving a relatively sharp maximum; in contrast, 2θ and χ scans barely affect the Bragg condition and so give relatively broad maxima. The lower limit on the width of the reflection profiles is set by the X-ray optical characteristics of the diffractometer (beam divergence) but this width will be increased markedly for poorer crystals with large mosaic spread. The exact hardware (half-shutters or diagonal slits) or software (iterative or single-pass centring) varies, but it is worth keeping some points in mind. First, if the half-shutters or slits are not correctly adjusted (see below), then reflection centring will at best be inefficient and at worst produce unreliable values for centred angles. Second, if your software does not iteratively centre reflections (until consecutive centrings produce angles in adequate agreement) you may find that you have to do this manually before you can achieve successful and reliable indexing. Third, make sure that any software settings such as those for collimator diameter or slit aperture match the settings on the hardware. There may be a set of parameters which control reflection centring, for example the agreement required between successive centring cycles: it is a waste of time demanding agreement to within 0.01° where reflections are very wide or poorly shaped, but too lax a requirement may mean that reflections are not as well centred as they could be, resulting in a poor orientation matrix. It may also be possible to specify the stepping intervals on the various axes during centring, but normally the centring routine will adjust these appropriately. If centring fails for no obvious reason, it is worth checking that these have not assumed ridiculous values (like $\Delta 2\theta = 0.00°$ or $\Delta \chi = 10.00°$!). If the χ (or κ) circle moves over a wide range during centring, this may indicate streaked reflections; if you have not already taken a photograph, you should do so. Unless you really know the unit cell, you should accumulate a list of at least 12–15 reflections before indexing (also see below).

5.3.2   Indexing, orientation matrix and cell determination

The mathematics of indexing have already been covered in Chapter 4, and apply essentially to all types of diffractometer. The aim is to find the simplest cell and matrix which allow integral indices to be assigned to all reflections. The purpose of this section is to cover the practical aspects, including possible strategies in cases where the first attempt fails.

Before starting the indexing routine, you should examine the indexing parameters; these place limits on the indices which can be assigned, on the lengths of the axes of the original primitive cell, and on the tolerances of the indices (how far they are allowed to deviate from integral values). Part of their original purpose was to reduce the computer time needed for indexing and to exclude ridiculously large cells, but the former is unlikely to be relevant to modern control computers and the latter is hard to judge unless you know the cell beforehand. It may be a good idea to start off with upper and lower limits on cell axes of 3 Å and 50 Å, respectively, and indices as high as 15 or 20 (assuming you have used the 2θ limits suggested above). If indexing fails, the first thing to try is increasing the limits on indices and axial lengths.

If indexing is unsuccessful, you need to examine the reflection list. Are there any reflections which are very weak or very strong, or have anomalous scan widths? If there are a small number of weak reflections, leave them out and try again. If indexing is successful, you may find that you can subsequently index these. If not, remember that any cell which fails to account for more than one or two weak reflections should be treated with great caution. Obviously, if a strong reflection fails to index, then the cell is most probably invalid, unless the crystal is twinned. This approach depends on having enough centred reflections so that a sufficient number remain for indexing after any suspicious ones have been omitted. Beware of crystals which give reflections whose angles match closely: this may indicate that the crystal has more than one component.

If you are still having problems and have not taken a photograph, this might now be advisable in case the crystal is visibly multiple. This can be more efficient than trying to manipulate an intractable list of reflections. (It is possible to obtain a matrix, either manually or with the aid of computer programs, and collect data on split or twinned crystals, but this is only worth considering if there is no better crystal.) If there is no obvious problem, you should eliminate the possibility that the crystal has moved during the search procedure, thereby giving a list of incompatible reflection positions. To do this, simply re-centre the first reflection; if it has moved, so has the crystal, and there is little alternative but to remount it securely, discard the reflections already found, and restart the search. If you suspect an instrument fault, you should check your circle zeros and re-centre the list of reflections before again trying to index. If such an error did occur, it was perhaps random and will not recur identically.

Even if indexing seems to have succeeded, you should be wary of unexpectedly large (often triclinic) cells with large uncertainties on their cell dimensions; if the cell is allowed to be large enough, it will be able to accommodate (approximately) almost any list of reflections. There is also a danger of choosing the wrong cell if the reflections used are too few in number (see above) or represent some subset. The most obvious cause of this is the presence of heavy atoms contributing to only some of the data. For example, a platinum atom on an inversion centre in $P2_1/c$ contributes only to reflections having $k + l$ even. These data will be stronger than the others and therefore more easily located, and if insufficient reflections with $k + l$ odd are located, the lattice type will be incorrectly assigned as A rather than P. Similarly, if only reflections with h even are included for indexing, the a axis will be half the true length and half the data set will not be collected.

If indexing appears to succeed, but one or more reflections have indices at or near special values such as $\frac{1}{2}$ or $\frac{1}{3}$, this may indicate the cell is too small in that direction and should be increased by a factor of two or three, respectively. If you decide to do this, you must take great care to ensure that this is justified, otherwise you will spend a lot of time measuring nothing; if you increase a cell axis length by a factor N, make sure that there are significant intensities for a reasonable proportion of reflections with corresponding indices which are not a multiple of N. An alternative and possibly safer approach is to search for and centre some more reflections: if the true cell is larger than the one first found, re-indexing should then yield it without the need to adjust the cell manually.

5.3.3 Finding the correct cell

Once the original primitive cell is found, the next step is to establish the correct Bravais lattice (crystal system + lattice type). This is done on metric considerations alone: for example, a transformed unit cell which has $a \neq b \neq c$, $\alpha = \gamma = 90°$ and $\beta \neq 90°$ (within the probably rather approximate standard uncertainties on these parameters) will be identified as monoclinic. Metric symmetry can be higher than Laue (diffraction) symmetry, and it is therefore important to check that reflections which are equivalent under the chosen Laue group actually have similar intensities. For a monoclinic crystal, the reflections hkl, $h\bar{k}l$, $\overline{hkl}$ and $\bar{h}k\bar{l}$ should be of similar intensity. This comparison is more difficult where strong absorption is present, but it is usually possible to make a clear decision. If all else fails, adopt the lower symmetry for data collection. If you are wrong you will get equivalent data to merge, but if you incorrectly choose the higher symmetry you may only have half the data you need. A common case is distinguishing between monoclinic and orthorhombic crystal systems, because metrically these differ only in that the latter has $\beta = 90°$, while in the former β can have any value, including ones which are 90° within standard uncertainty. These systems differ in their diffraction symmetry in that, for the orthorhombic case, all permutations of $\pm h$, $\pm k$, $\pm l$ give equivalent reflections and their intensities can be measured and compared. Further difficulties arise with higher symmetry crystal systems where there is no one-to-one correspondence between crystal system and Laue group; for example, a tetragonal crystal may have either 4/*m* or 4/*mmm* Laue symmetry, but the solution is the same: measure a number of sets of reflections which should have the same intensity under each Laue group and see whether they agree. Again, if in any doubt, collect data in the lower symmetry Laue group.

There may still be doubt as to whether the unit cell is correct, and there are two approaches to checking this. The first is to record axial photographs. Each of these is taken over a limited oscillation range about a cell axis or other direction, in order to reveal axial length and symmetry information. The second is to multiply all three axes by a certain factor (e.g. N) and then perform a high-speed data collection to see whether any reflections with an index which is not a multiple of N have significant intensity; if not, the original cell (i.e. without the multiplied axes) can then be used with more confidence. If an ancillary symmetry-checking program is available, use it.

At this stage some other, external, checks are a good idea. For example, there is probably little point in continuing if the unit cell has been reported before, either in the literature or within your own department. To exclude the first possibility, the Crystal Data Information File (CDIF) should be consulted. This contains information on around 240 000 reported unit cells. It is available to UK academic researchers via the Chemical Database Service at CLRC Daresbury Laboratory (e-mail: uig@dl.ac.uk), and similar arrangements may apply in other countries. Alternatively, it may be supplied as part of your diffractometer's software suite; if so, make sure you get regular updates. To guard against repeating in-house determinations, you should keep your own database of unit cells. A positive match at this stage is slightly dispiriting, but you will have invested only an hour or two. If undetected until after data collection and structure solution, such a duplication could easily waste several days.

Another useful check is whether the cell volume is compatible with the molecular formula of the expected compound, allowing $18 \, \text{Å}^3$ per non-hydrogen atom. This value is surprisingly valid for a wide range of organic, metal–organic and coordination compounds, although some adjustment might be required for special cases (e.g. $14 \, \text{Å}^3$ for some highly condensed aromatic compounds, $22 \, \text{Å}^3$ for some organosilicon compounds). It should not be applied to purely inorganic compounds. A significant discrepancy may indicate that the cell is wrong, that the compound is not as proposed, or that solvent molecules are present. The number of molecules present in the unit cell may forewarn you that, if the proposed molecular formula is correct, disorder must be present.

5.3.4   Obtaining a good orientation matrix

The initial matrix and cell, especially if determined on a four-circle diffractometer using low angle reflections from a rotation photograph, will not be sufficiently precise for data collection. Reflections with higher Bragg angles are required, so that the relative errors in setting angles are reduced. Values of 2θ around 30° for molybdenum and 50° for copper radiation are high enough to give good resolution without incurring problems due to splitting of the α_1–α_2 doublet, although some instruments can cope with this. The fall-off of intensity at higher angles can be a problem with weakly diffracting crystals, and some compromise may be necessary by reducing the 2θ range. Peak-top measurements or a rapid data collection within the chosen range should identify sufficiently intense reflections from which a suitable number, perhaps between 12 and 25, are selected. This list can be compiled in different ways: (i) all the reflections are independent and belong to the unique set(s) of data which are to be collected; (ii) a small number of reflections and their symmetry equivalents are chosen; (iii) the required number of reflections is chosen from the complete sphere of data. Each of these has its own advantages and drawbacks. Reflections should be chosen with $\sum |h|/a \approx \sum |k|/b \approx \sum |l|/c$ so that the precision of the matrix (and of the derived cell parameters) is as similar as possible in all directions; this is particularly important in (ii) above. At this stage it is probably time to set the hardware and software parameters for the slit apertures to the values to be used in data collection. Once the list has been centred, perhaps using manual iteration, refinement should yield an orientation matrix suitable for data collection. If the matrix is significantly different or better determined, a re-check of the unit cell and symmetry is recommended.

5.3.5   Obtaining the best unit cell dimensions

The unit cell obtained from the refinement of the orientation matrix is not the most suitable for use in structure analysis and for final reporting of results. It is based on a limited number of centred reflections, the cell is refined without symmetry constraints (it is effectively treated as triclinic regardless of the true crystal system), and it is prey to systematic errors in angular positions resulting from diffractometer zero errors and crystal mis-centring. Cell parameters are best determined after data collection, by a symmetry-constrained refinement based on 2θ values only, these being derived by measurements on either side of the X-ray beam ($2\theta = \omega_+ - \omega_-$) on a much larger number of reflections selected from the full data set. At the very least the uncertainties on unit cell parameters should be sufficiently low that they do not make a significant contribution to the uncertainties in the molecular parameters derived from structure refinement.

5.4   Preparing for data collection

5.4.1   Introduction

Once started, data collection is a fully automated procedure, so it is particularly important that the parameters controlling it are optimally set before commencing. Getting these parameters wrong can result in anything from inefficient use of instrument time to a completely useless data set. The aim of the experiment is the measurement of intensity data to the maximum possible precision in the minimum time, while avoiding errors which could degrade the quality of the data set. The most obvious constraints are imposed by the quality of the available crystals (size, shape, diffracting power, reflection profiles, fall-off in intensity with increasing θ, stability), but the diffractometer must also be considered (its general reliability, reproducibility of circle positions, stability of the X-ray tube intensity, the settings and stability of its counter and related electronic components, mechanical restrictions on the circle positions, and the suitability and power of the control software). Finally the user has to take certain decisions such as the time allocated to each reflection (and consequently to the whole collection procedure), how this is partitioned between reflections of different intensity, the scan type and width to be used, and the frequency with which standard reflections are checked and reorientation carried out. The purpose of a particular experiment will have a bearing on some of these decisions.

5.4.2 Parameters

(i) Scan type and width

Scanning a reflection usually involves movement of the crystal (ω) with possibly some movement of the detector (2θ). A pure ω scan (no detector movement) reduces the overlap between neighbouring reflections and is advisable where long cell axes or wide reflection profiles are present, and particularly when using molybdenum radiation. An $\omega/2\theta$ scan (the detector moves at twice the angular rate of the crystal) is necessary to record reflections when the mosaic spread is very low. In general, an ω/θ scan (crystal and detector move at the same rate) is often a good compromise and on some diffractometers it allows faster data collection than $\omega/2\theta$ scans. The scan type must therefore be chosen to suit the crystal being studied. The scan width can be difficult to judge, more so if the profile width changes with crystal orientation, the peaks are not in the centre of the scan, or the crystal moves during data collection. Setting too wide a scan will waste diffractometer time, but too narrow a scan is worse, because it will truncate reflections and introduce systematic errors into your data. Where profile width changes markedly with crystal orientation, control software which allows for this is especially valuable, as an appropriate scan width is calculated for and applied to every reflection.

(ii) Detector aperture

This should be narrow enough to reduce background radiation and thereby increase precision, but wide enough to avoid truncating reflections, which would introduce systematic errors. As there is correlation with the collimator diameter, scan type and scan width, the aperture should be set using the same conditions as will pertain during data collection. The aperture can be reduced stepwise and a small number of representative reflections scanned to ensure that the reflections are not being truncated.

(iii) Collection speed

The speed at which you can safely collect diffraction data will depend on the intensity of your X-ray source (synchrotron $\gg$ rotating anode $>$ sealed tube), but for a particular configuration it will be determined by the diffraction quality of the crystal under study. There is little point collecting data so rapidly that most of your data are weak. Conversely, excessive data collection times are wasteful. If you intend to solve your structure by direct methods, at least half the data between 1.1 and 1.2 Å resolution should be 'observed'. This corresponds to approximate 2θ ranges of 34–38° for molybdenum and 80–90° for copper radiation, and a short data collection in the appropriate shell should tell you whether your choice of speed is about right. Different collection rates may be appropriate for different 2θ shells.

(iv) Weak reflections

These are neither insignificant nor useless, and can be valuable in both solving and refining structures: the problem is how to manage their measurement. One method uses 'pre-scans', in which each reflection is scanned relatively rapidly, before a triage is carried out by comparing the resulting $I/\sigma(I)$ ratio with pre-set upper and lower thresholds. A reflection which exceeds the upper threshold is deemed to be 'strong' and the measurement is accepted; one which does not reach the lower threshold is deemed 'weak' and no further measurement is made; those between the two thresholds are scanned for an estimated additional time in an attempt to reach the upper threshold, subject to a time limit per reflection. This method has the advantage of flexibility, but the resetting of circles and rescanning of reflections which have to be measured twice will reduce efficiency. Choosing the various counting times and threshold parameters requires some skill and care. The main drawback is that weak reflections can be very poorly estimated. An alternative method, which is more compatible with modern direct methods and with refinement against all data, is to spend the same time (more or less) measuring each reflection. With most control software, this requires setting the counting times per reflection (or scan step), and the upper and lower $I/\sigma(I)$ thresholds, to be the same. Although this method is simpler to set up, since only the counting time is important, it lacks the safety net inherent in the triage method, where the diffractometer software makes decisions about which reflections to rescan. The bonus is that no resetting or rescanning is involved and the collection is more efficient. Profile-fitting or profile analysis routines should be used wherever available, as they will enhance the precision of the weak reflections, but they may not be able to cope reliably if the reflection profiles show serious splitting, and should probably be avoided in such cases.

(v) Intensity standards: checking and reorientation

Standard reflections play a vital role during data collection in monitoring the crystal for movement or decay. They should be well distributed in reciprocal space, give strong to medium intensity, and have intermediate 2θ values. Although the reflections must be reasonably intense in order to achieve high $I/\sigma(I)$ values, avoid strong reflections at low angle, as these are the most likely to change due to an irradiation-induced increase in mosaic spread. At least three reflections should be used: these should be approximately orthogonal to each other (axial reflections may be useful) in order to be sensitive to crystal movement in all directions. The frequency with which standards are monitored will depend on the experiment, but it might be reasonable to do so at least once every two hours for ambient measurements, and every hour for low-temperature work. If any of the standards have dropped in intensity or moved (by more than the limits set), or if a set time has passed (or a predetermined number of standard measurements has been made), the crystal will be reoriented by re-centring a list of reflections. These parameters (e.g. a 10% drop of intensity or a shift of 15% of the scan width) need to be tailored to the particular case. Remember that reorientation can take considerable time if a long list is used, and it involves extensive driving of the circles, which may aggravate crystal movement. Pre-emptive re-centring should be carried out only two or three times in 24 hours unless there is a serious problem with crystal movement. If

your crystal does suffer decay, the standard measurements provide a basis for applying
a correction.

(vi)   Choosing which reflections to measure
The time required for data collection depends strongly on the upper 2θ limit. The number
of reflections to be collected is proportional to $\sin^3\theta$: collecting a data set to $2\theta_{max} = 60°$
will give more than twice as many reflections as to $2\theta_{max} = 45°$. It is worth noting
the minimum upper limits on 2θ recommended by *Acta Crystallographica* (Mo 50°,
Cu 134°, both equivalent to a d-spacing of 0.84 Å). These limits are useful guides, but
not all crystals diffract to these limits and there is no point in trying to collect useful
data where reflection intensity has decreased effectively to zero. Compounds containing
heavy atoms often diffract well to high angles, but it is largely a waste of time setting
a very high value for $2\theta_{max}$: the high-angle data are dominated almost exclusively by
contributions from the heavy atoms and will not help with the much more difficult task
of accurately locating the lighter atoms. High-angle data are of greater value in cases
of anomalous dispersion (e.g. establishing absolute configuration) since, unlike normal
scattering which falls off with increasing θ, $\Delta f''$ is independent of θ.

Unless the control program has retained the information from the original Bravais
lattice determination, it will be necessary to enter the lattice type as a data collection
parameter. As lattice centring absences cause at least half the possible data to be systematically absent, this represents an important saving in time, but you must be sure
that the lattice type is correct. It is not worthwhile setting absence conditions for glide
planes or screw axes, as relatively few reflections are affected and any gain is wholly
disproportionate to the consequences of an error.

The unique set of data for triclinic, monoclinic and orthorhombic crystals comprises
a hemisphere, a quadrant and an octant, respectively, although Friedel opposites should
also be measured in non-centrosymmetric cases and may be essential if an absolute
structure determination is required. In a given time, it is possible to measure either the
unique set relatively slowly, or two or more symmetry-equivalent sets more quickly.
Both methods have advantages. Collecting just the unique set means that relatively less
time is spent driving between reflections and is likely to give rise to a higher proportion
of 'observed' data. However, the collection of symmetry equivalents allows merging,
which can reduce systematic errors (and indicate their severity) as well as providing a
check on internal consistency. The faster data collection may yield a complete unique
set in cases of crystal decomposition or equipment failure. If the collection of the first
unique set of data is going to finish in the absence of an operator, the instrument should
be set to begin a second set in order to take full advantage of the time available (and
perhaps prevent icing in low temperature collections). A compromise which achieves
some of the advantages of both methods is to collect somewhat more than a unique set:
for example, while the unique set for an orthorhombic crystal might be defined by the
index ranges $h\ 0 \rightarrow 10$, $k\ 0 \rightarrow 12$, $l\ 0 \rightarrow 17$, substituting -1 for 0 as the lower
index limit on each axis would allow useful checks on diffraction symmetry and internal
consistency without adding greatly to the data collection time. However, you should
always be cautious in your interpretation of merging R values when the second unique
set is incomplete.

If the control program does not automatically select the most efficient order in which to collect data, it will allow you to control this manually. In general, for a primitive lattice at least, the index which changes fastest should correspond to the longest cell axis (and the slowest to the shortest). This will minimize the time spent driving between reflections, especially if successive rows of reflections can be measured in zigzag order.

(vii) Final checks

You are now ready to initiate the automatic data collection. Before you begin, you may want to take the precaution of re-centring the first reflection found by your search procedure; if it has not moved, that provides reassurance that the crystal is stable. If you have not yet carried out a ψ scan, it may be instructive to do so: an absorption curve which is inconsistent with the crystal morphology and proposed composition may cause you to re-evaluate whether data should be collected. For example, if a plate-like crystal reputed to contain one or more strongly absorbing elements gives an essentially flat ψ scan, you should be very suspicious: the crystal could be of a ligand rather than the expected metal–ligand complex. Less commonly, you may find that the crystal shows much stronger absorption effects than expected and that you must make more extensive measurements than you had anticipated. Finally, you should check that the X-ray generator settings are correct, that the flow of cooling water is adequate and stable, and that any low-temperature device has sufficient cryogenic fluid.

5.5 Data collection

The procedure of data collection is essentially automatic, but it is advisable to be present while the diffractometer carries out any checks (e.g. on its zero positions) and moves on to measure the standard reflections. Check that each of these has sufficient intensity and that its maximum lies near the centre of a scan of adequate width. If practical, the data collection should be checked after the first remeasurement of standards, in case there has been crystal decay or movement.

The data collection routine may maintain a summary of the numbers of reflections which are inaccessible (due to hardware restrictions), are 'unobserved', lie off-centre, have uneven backgrounds, or show scan width errors. If a substantial proportion show one or more of the last three symptoms, manual intervention may be warranted, but if the procedures for setting up the crystal described above were applied, this is unlikely to be necessary or result in any improvement. The degree of tolerance shown towards such symptoms will depend on your assessment of the quality of the crystal which you obtained during setting-up: a good crystal showing such effects is particularly suspicious and must be investigated. Some control programs display reflection profiles or build up a plot of the intensity-weighted reciprocal lattice as the data are collected, and an examination of these can be extremely valuable in identifying possible problems.

If, after several hundred general ($h\,k\,l$) data have been collected, the proportion of observed data is less than one-half, it is worth checking whether there is any pattern to the absences. Has lattice centring been overlooked, or is a cell axis a multiple of its correct value? If either of these is true, remedial action is required. This procedure must not involve axial or zonal reflections, as these can be affected by absences due to screw axes and glide planes, respectively. Even if there are no problems, you should check on the data collection regularly to ensure all is going well.

5.6 Gross systematic errors

Systematic errors due to factors such as absorption and crystal decay are often predictable and largely amenable to correction using standard procedures. The same cannot be said for the totally false measurements caused by instrument misalignment or failures such as shutters or attenuation filters sticking (usually in the fully open or closed positions), generator problems, electronic instabilities, mains spikes, or incorrect attenuation filter factors. The only way to avoid these is by regular maintenance and permanent vigilance. For example, problems with attenuation filters may show up after structure refinement as large disparities between individual F_o^2 and F_c^2 values; source instability may be detectable by examining the variation in intensity standards, especially if the crystal itself is well-behaved. Regular checking of the instrument alignment with a standard crystal is essential. While the crystal is there, check the unit cell: are the cell parameters truly accurate? If you have the time to collect some data, make sure you collect plenty of equivalents and ensure that their intensities are in good agreement. How good is the structure obtained by refinement against such data? Are there any large disparities between individual F_o^2 and F_c^2 values? If there are problems with your test crystal, you must identify the source of these and take remedial action. If, during normal operation, a similar problem seems to afflict a series of crystals, especially if these are of very different materials, you should investigate whether the problem is instrumental.

5.7 Correction of intensity data

The process of data reduction requires the application of a number of corrections. Some of these (Lorentz and polarization corrections) must be applied to every set of data and, provided that your data reduction program is using the appropriate correction factors, no operator input or intervention is needed. Other corrections such as those for absorption do not apply in every case, but you can predict whether they are necessary if you know the composition of the crystal. There may be no warning of the need for substantial crystal decay corrections until the intensity standards begin to fall.

5.7.1 Absorption corrections

As a rough guide, any data set collected with copper radiation may require absorption correction, as will any molybdenum data set from a crystal containing elements heavier than about silicon. The severity of absorption effects will depend on how many of the heaviest atom are present and on the crystal morphology. More precisely, the need for an absorption correction can be gauged by calculating μx, where μ is the linear absorption coefficient (in mm^{-1}) corresponding to the (assumed) cell contents and the radiation used, and x is the average crystal dimension; if μx is less than about 0.1, then no correction is necessary. The differential effects of absorption may be estimated by calculating $e^{-\mu x}$ for the most extreme crystal dimensions.

There are two crystal-based methods for making absorption corrections. The first, numerical integration, is based on accurate indexing of crystal faces and measurement of their distance from a common reference point within the crystal. The set of faces must define an enclosed crystal volume. This method is the most rigorous and strongly recommended in cases of strong absorption. It is obviously inapplicable where the crystal lacks identifiable faces or when the faces or dimensions are ill-defined because the crystal is protected in a film of oil or within a capillary tube. However, there are some computer programs which attempt to apply modified corrections starting from an approximate description of the crystal. It is obviously important (a) that the indices of the faces are correctly identified, and (b) that the distances are accurately measured. There may be software support here, for example by drawing an illustration of the crystal for comparison with the view seen down the microscope.

The second crystal-based method, which requires less operator effort but more diffractometer time, is based on azimuthal (ψ) scans, where the crystal is rotated about the scattering vectors of a number of reflections. As a reflection remains in a diffracting position throughout, the variation of its intensity as a function of crystal orientation can be measured (Fig. 5.2). At $\chi = 90°$ the movement corresponds to rotation about ϕ, at other χ values to a more complicated combination of circle movements. It is important to perform ψ scans on reflections with a range of 2θ values (in order to model the dependence of absorption on the Bragg angle) and χ values, although the range of the latter may be limited by low-temperature attachments or by circle geometry (but absorption correction routines should be able to utilize partial scans). The ψ increments used must be appropriate: they should be small if absorption changes markedly over a narrow angular range, as will happen for a platy crystal with a high value of μ. If the crystal is more ellipsoidal, or μ is smaller, you could measure more reflections with a larger ψ increment. Because they are going to be used to apply corrections to the whole

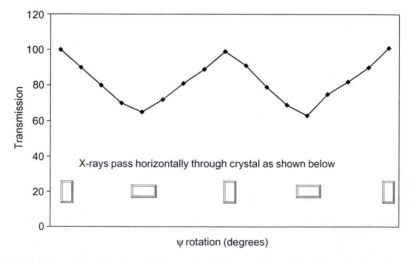

Fig. 5.2. Variation of intensity with angle in an azimuthal scan for a crystal with rectangular cross-section.

data set, ψ scans must be examined very critically to ensure that, as far as possible, they represent only absorption effects. If a reflection is split at some angles but not at others, if the width varies with ψ so that it is truncated at some positions, or if measurements at ψ values 180° apart disagree badly, it should not be used. Variations in intensity which are incompatible with calculated values of $e^{-\mu x}$ should be regarded with suspicion, although until the structure is solved you cannot be sure whether the assumed value of μ is correct. If a structure turns out to be different from the one expected and the previously assumed value of μ was therefore significantly in error, you will need to reprocess the data using the correct value, regardless of whether face-indexing or ψ scans have been used.

Empirical methods which require no special measurements can be applied once the structure has been solved and refined to isotropic convergence. They work by calculating an absorption surface to minimize the sum of the weighted differences between observed and calculated structure factors (or their squares). If neither face-indexing nor ψ-scan methods can be applied, these empirical methods represent a last resort, but are the subject of continuing controversy: their use in programs such as *DIFABS* and *XABS* has been criticized on the grounds that the derived absorption corrections are biased, that they include corrections for systematic errors other than absorption, and that they are often used inappropriately. If you have to use such methods, make sure that the program does effect an improvement, by monitoring its internal residuals and checking that the absorption correction range is reasonable. Use of this method is incompatible with certain types of work: it distorts the anisotropic displacement parameters and you cannot, for example, derive meaningful librational corrections for molecular geometry parameters.

5.7.2   Decay corrections

If a crystal moves significantly during data collection, this should be detected by changes in the standard reflections and appropriate remedial action taken. Once the data collection is complete, the standard measurements can be used to apply corrections for crystal decay (or increase in mosaic spread) or for instrumental instabilities (provided that these occur on a similar time-scale to the standard measurements). The standard measurements can be applied in various ways, such as pairwise linear interpolation between sets of standard measurements, batch scaling (useful if part of your crystal has split off during data collection), and polynomial curve-fitting. The general assumption of isotropic decay is usually valid, but can be checked by looking at the variation in the individual standards. The purpose of a decay correction is to place intensity measurements taken at different stages of the data collection on the same basis. For example, if the standards have fallen to exactly half their initial values by the end of the data collection and a linear correction is applied, the last reflection collected would be multiplied by a factor double that for the first one.

5.7.3   Other possible corrections

Extinction predominantly affects strong, low-angle reflections, and is normally corrected for by including a single correction factor as a variable in the structure refinement.

Secondary extinction is wavelength-dependent, being worse with copper than with molybdenum radiation.

Thermal diffuse scattering (TDS) can artificially enhance the intensity of some high-angle reflections. The fact that TDS decreases with temperature provides yet another incentive to collect low-temperature data.

Multiple diffraction effects are more likely to occur if a prominent lattice vector is aligned with the ϕ axis. They are most obvious where they cause significant intensity to appear at the position of a systematic absence. If their significance is not noted, they can cause problems with space group determination. They can also be recognized by their anomalously narrow reflection profiles.

Some data reduction programs will attempt to compensate for the effects of crystals which are larger than the X-ray beam: the collimator diameter and the crystal dimensions and orientation are required. You are strongly recommended to avoid this situation rather than trying to correct for it.

Exercises

5.1 Assuming both are available, which of Cu or Mo radiation would you use to determine the following structures, and why:
(a) $C_6H_4Br_2$; (b) $C_6Cl_4Br_2$; (c) $C_{36}H_{12}O_{18}Ru_6$; (d) absolute configuration of $C_{24}H_{42}N_2O_8$; (e) absolute configuration of $C_{24}H_{40}Br_2N_2O_8$?

5.2 You have a crystal of a platinum complex, which is weakly diffracting. What are your options?

5.3 It has not been possible to index the following list of centred reflections and obtain a unit cell and crystal orientation matrix. Suggest possible reasons for this failure.

N	2θ	ω	χ	ϕ	Intensity	Width ($°$)
1	12.23	6.50	35.42	157.62	971	1.37
2	12.42	6.43	32.61	69.52	1217	1.57
3	11.41	0.13	−70.93	26.24	62	2.28
4	−9.46	0.15	−1.77	29.92	52 117	1.02
5	10.91	0.13	32.62	56.82	1347	1.52
6	−11.91	0.14	−57.67	57.35	923	1.29
7	10.56	0.07	56.29	4.47	9986	0.50
8	11.14	0.12	60.53	11.29	953	1.32
9	−10.96	0.19	32.67	43.67	1323	1.56
10	16.82	8.63	−54.54	86.95	439	4.19
11	17.40	8.60	−53.94	174.75	66	3.34
12	18.29	9.33	26.34	64.00	563	1.25
13	16.07	7.94	−28.28	87.62	1053	1.18
14	15.13	7.84	−28.57	134.61	392	1.24
15	18.26	9.27	26.40	64.06	457	1.59

5.4 A crystal indexed to give a metrically orthorhombic unit cell. The following were part of a more extensive series of measurements made to check the diffraction symmetry. There were no major absorption effects. Is the crystal really orthorhombic?

h	k	l	Intensity
10	2	4	258.2
−10	2	4	187.4
10	−2	4	267.0
10	2	−4	216.4
−10	−2	−4	245.2
10	−2	−4	200.9
−10	2	−4	264.6
−10	−2	4	208.3

5.5 A compound $C_{32}H_{31}N_3O_2$, which has been crystallized from tetrahydrofuran (C_4H_8O, THF) solution, gives a primitive monoclinic unit cell of volume 1850 $Å^3$. What are the likely unit cell contents?

5.6 Assess whether the reflections $(3, 9, 10)$, $(1, 5, 11)$, $(−2, 4, 9)$, $(−3, 10, 1)$, $(−1, 3, 8)$, $(0, 4, 14)$ and $(2, 4, 8)$, together with their symmetry equivalents, are suitable for refining the orientation matrix corresponding to a C-centred monoclinic cell with $a = 5.65$, $b = 15.92$, $c = 10.24$ Å, $\beta = 99.6°$.

5.7 For the same unit cell as in Exercise 5.6, which, if any, of the following sets of reflections would be suitable as intensity standards?

	h	k	l	2θ	Intensity
(a)	−6	0	0	45.1	384
	0	6	0	15.5	25 064
	0	0	8	32.7	1576
(b)	−4	0	0	29.6	3005
	0	8	4	26.3	3567
	−2	6	8	37.2	1724
(c)	−4	0	0	29.6	3005
	0	10	2	27.0	589
	1	1	7	30.8	164

5.8 Estimate the range of absorption correction factors expected for the following crystals with $\mu = 1.0$ mm^{-1}:
(a) a thin plate 0.02 × 0.4 × 0.4 mm;
(b) a tabular crystal 0.2 × 0.4 × 0.4 mm;
(c) a vertically mounted needle 0.06 × 0.08 × 0.40 mm;
(d) a horizontally mounted needle 0.06 × 0.08 × 0.40 mm.
Repeat the calculations with $\mu = 0.1$ and with $\mu = 5.0$ mm^{-1}.

5.9   Two estimates were made of a set of unit cell parameters:
       (a) $a = 8.364(12)$, $b = 10.624(16)$, $c = 16.76(5)$ Å, $\alpha = 89.61(8)$, $\beta = 90.24(8)$,
           $\gamma = 90.08(6)°$
       (b) $a = 8.327(4)$, $b = 10.622(6)$, $c = 16.804(8)$ Å, $\alpha = 90$, $\beta = 90$, $\gamma = 90°$.
       The first was derived from orientation matrix refinement on centring angles for 12 reflec-
       tions, the second from 62 pairs of measurements of 2θ at $\pm\omega$. Estimate the approximate
       contribution in each case to the uncertainty in a C–C bond of 1.520 Å.

5.10   In data collection, it is normal to spend twice as long measuring the reflection peak as is
       spent measuring the aggregate of the left background (LB) and right background (RB). The
       net intensity of a peak is therefore given by

$$I_{net} = I_{peak} - 2(I_{LB} + I_{RB})$$

and its uncertainty by

$$\sigma(I)_{net} = [I_{peak} + 4(I_{LB} + I_{RB})]^{1/2}.$$

Calculate I_{net} and $\sigma(I)_{net}$ for the following reflections:

hkl	I_{peak}	I_{LB}	I_{RB}
(a) (2, 6, 2)	560	55	85
(b) (−1, 4, 5)	2304	70	74
(c) (0, 11, 0)	140	42	35
(d) (0, 24, 0)	120	16	18

6

Area detectors

6.1 Introduction

Until a few years ago, the vast majority of single-crystal diffraction data collection
for chemical applications was carried out with four-circle diffractometers. Electronic
area detectors found some limited application, most notably in the EPSRC-funded UK
National Crystallography Service, then at Cardiff, but the main users of area detectors were protein crystallographers. The introduction of a few image-plate systems into
chemical crystallography laboratories in the early 1990s was followed in 1994 by the first
available commercial charge-coupled device (CCD) detectors incorporated into complete
turnkey laboratory systems. In the intervening years these have become widely available
from a number of suppliers, have dropped in price, and have established themselves as
the technique of choice among chemical crystallographers. They are rapidly displacing
four-circle diffractometers as the standard workhorse data collection equipment.

The underlying theory for data collection using area detectors is the same as for four-
circle diffractometers, being based on the concept of the reciprocal lattice and the Bragg
equation. The crystal orientation matrix, already discussed, still plays the same central
role in the calculation of diffraction conditions and diffractometer geometry settings,
but the detailed algebraic applications are different. The replacement of a single-point
detector by an area detector brings with it both advantages and disadvantages, though
the advantages are generally regarded as being overwhelming. They are summarized in
Table 6.1.

6.2 Types of area detectors

One form of area detector has, of course, been used in X-ray crystallography from the
beginning of the subject: photographic film. This has many desirable properties, but
suffers from the disadvantages that it requires chemical development of the image, it is
not reusable, and there is no possibility of knowing just when a particular incident X-ray
beam is recorded (no time resolution). It also does not have a very high dynamic range (the
range of intensities which can be reliably recorded). Most forms of X-ray photograph
rely on having an oriented crystal, for which one unit cell axis or other major lattice
vector is aligned in a particular direction (such as along or perpendicular to the incident
X-ray beam). Although this is not necessary in order to observe diffraction effects, it
does produce more easily interpreted diffraction patterns, the precise details depending
on the geometry of the particular type of X-ray camera used (rotation, Weissenberg,
precession, etc.). Film techniques are increasingly regarded now as of mainly historical
interest.

Table 6.1   Area detectors versus conventional diffractometers

Advantages
- Simultaneous recording of many reflections
- Much faster data collection possible
- Data collection time independent of structure size
- High redundancy of symmetry-equivalent data possible
- Rapid screening of samples
- Not necessary to obtain correct unit cell before data collection
- Complete diffraction pattern measured, not just Bragg reflection positions
- Reduced probability of obtaining incorrect cell from few initially found reflections
- Poorer crystal quality and weaker diffraction can often be tolerated
- Less crystal movement necessary; easier to use low-temperature and other accessories
- Easier visualization of the diffraction pattern, good for teaching and training

Disadvantages
- Possibly higher capital and maintenance costs
- Requirement for high-speed and high-capacity computing
- Need for careful corrections for non-uniformities and other effects
- Usually less discrimination against other X-ray wavelengths, e.g. harmonics
- Restricted detector size may lead to difficulties with large unit cells and with Cu Kα X-rays

Modern area detectors are effectively electronic versions of photographic film, which overcome some or all of its disadvantages to various extents. They do, however, have their own disadvantages, which may include, for different types of detector, high cost and limitations on physical size, dynamic range of recording, sensitivity, spatial and temporal resolution. The continuing improvement of existing types of area detectors and the development of new types, some of this with major public and commercial funding, is rapidly reducing these disadvantages.

The major obvious advantage of electronic area detectors over conventional scintillation counters on diffractometers is their ability to record diffraction data over a substantial solid angle. As far as normal Bragg diffraction is concerned, this means the simultaneous measurement of a number of reflections instead of a serial measurement of reflections in some order; the number of reflections measured simultaneously depends on the size of the unit cell as well as on the size of the detector. In addition, however, an area detector actually records the whole of the intercepted diffraction pattern and not just the Bragg reflections, i.e. the whole of reciprocal space is observed, and not just the regions immediately around the reciprocal lattice points. This can be useful for special purposes, such as the study of incommensurate structures, but is not yet generally exploited in structure analysis. A related advantage is that it is not in fact necessary to establish the correct unit cell and crystal orientation matrix before beginning data collection (although there are good reasons for doing so whenever possible), because these can be found later from the recorded and stored images. In the case of a four-circle diffractometer, an incorrect orientation matrix means a useless set of data, because only the predicted reflection positions are explored.

On the other side of the balance, disadvantages of area detectors include not only the capital cost (by now a much less significant factor, as demand for area detectors grows and interest in four-circle machines wanes), but also the higher demand on computing resources to deal with larger data sets and faster measurement of them, and the need for careful corrections for non-uniformity of both the spatial and the intensity response of the detectors. A complete set of images for one crystal may require as much as a gigabyte or more of storage and considerable memory for processing to extract the reflections from it. The inexorable growth in power and decline in cost of computers means that this is also not a major concern any more.

There are several different types of area detector currently in use on X-ray diffractometers.

1. Multi-wire proportional chambers (MWPC). In proportional counters each incident X-ray photon causes an ionization in the detector gas, and hence a current in a high-potential wire. MWPCs usually have one set of parallel wires and a second set at right-angles, and a current is induced in one or more wires of each set for each X-ray photon. Thus both the time and position at which the photon arrives are known. The MWPC has the advantages over other area detectors that the output signal is instantaneous (time resolution is ideal), there is no inherent noise in the detector, and the counting efficiency is high. There is, however, a problem with parallax, particularly at shorter wavelengths, reducing the spatial precision, and the overall count rate is limited by the dead-time, which affects the detector as a whole for every single recorded photon. MWPCs have mainly been used up to now for protein crystallography with copper radiation, for which the parallax problem is not so important and intensities are not particularly high. They have not been used significantly for chemical crystallography. Detectors which can tolerate greater count rates are currently being developed, especially for use on synchrotron sources, but they are very expensive, both to build and to maintain.

2. Phosphor coupled to a TV camera. The coupling is by fibre optics, the phosphor converting incident X-rays to visible light for detection by the low-light-level TV camera. This type of detector also gives an instantaneous readout, but the active area is relatively small, as is the dynamic range, and the signal-to-noise ratio is less good than for other area detectors. The TV system most widely used is the Enraf-Nonius FAST; its application for chemical crystallography has been mainly in the EPSRC National Crystallography Service at Cardiff up to 1998. It is no longer commercially available, and is of historical interest only.

3. Image plates (IPs). Instead of converting X-rays to visible light, the phosphor in these devices stores the image in the form of trapped electron colour-centres. These can subsequently be 'read' by stimulation from laser visible light (which causes them to emit their own characteristic light for detection by a photomultiplier) and then erased by strong visible light before another exposure to X-rays. The main advantages of image plates are that they are available in large sizes and are relatively inexpensive; they also have a high recording efficiency and a high spatial resolution. The one major disadvantage is the need for a separate read-out process, which involves complex mechanics to scan the laser beam over the whole surface and is much slower than for any other electronic area detector, often taking minutes rather than seconds. Faster image plates

are currently being developed, and another method of reducing the time unavailable for X-ray exposure is the use of two or more plates, so that one is recording an image while another is being read, although this adds to the cost and complexity. Commercial IP systems used for chemical crystallography are supplied by a number of manufacturers; image plates can also be obtained from other suppliers and used with off-line scanning, but this is much less convenient.

   4. Charge-coupled devices (CCDs). This type of detector, more familiar in video cameras and other mass-market applications, is a semiconductor in which incident radiation produces electron-hole pairs; the electrons are trapped in potential wells and then read out as currents. Direct recording of X-rays is not usually carried out, for various reasons. Instead, a phosphor is fibre-optically coupled to the CCD chip, which is cooled to reduce its inherent electronic noise level (thermal excitation of electrons). The full sequence of events occurring is: (a) X-rays hit the phosphor at the front of the detector and produce visible light; (b) fibre optics transmit the light to the smaller CCD chip, usually demagnifying the image; (c) electron/hole pairs are generated within the chip by the incident light; (d) electrons trapped in potential wells are read out as currents after completion of the desired exposure time. Efficient recording, a high dynamic range and a low noise level, and a read-out time measured in seconds or fractions of a second, combine to give the CCD some clear advantages as a rapid area detector, but so far only relatively small good-quality chips are available, requiring substantial fibre-optic expansion to achieve even a moderate active phosphor area in most cases. Of the various types of area detector now in use, this has the best potential for further development, and considerable effort is being made in producing larger and more sensitive chips without significantly increasing the read-out time. Commercial CCD systems for chemical crystallography are now available from several major firms. The next few years are likely to see the introduction of so-called 'pixel arrays', another form of solid-state detector which should be able to record X-rays directly and over a larger area.

   Area detectors do not just offer the possibility of collecting diffraction data more quickly, although that is generally perceived as their main advantage. They also make feasible experiments which are beyond the scope of traditional four-circle diffractometers, through higher sensitivity and the recording of the whole pattern.

   Much of the following discussion applies equally well to any type of area detector, but we shall refer specifically to CCD systems, since these are most widely used in chemical crystallography.

6.3   Some characteristics of CCD area detector systems

An area detector, whether CCD or some other kind, is only one component of a single-crystal diffraction experiment. It needs to be combined with a goniometer for mounting and moving the crystal sample, a source of X-rays, and electronic and computing control systems. The detector itself consists of a face-plate and phosphor, usually a fibre-optic glass coupling from the phosphor to the CCD (though other forms of optical coupling are beginning to be used), the CCD chip itself with its electronics, and a cooling system. The details of these components vary with suppliers, though all are designed to do essentially the same job, and technical specifications continue to change.

Although an area detector records a number of diffracted beams simultaneously, it is still necessary to rotate the crystal in the X-ray beam in order to access all the available reflections. In most chemical crystallography systems, the detector is offset to one side rather than being held perpendicular to the incident X-ray beam, so that a higher maximum Bragg angle can be observed for a single detector position. Typical designs and configurations give data to a maximum Bragg angle θ of around 25–30°, appropriate to a 'standard' structure determination with a more than adequate data to parameter ratio if Mo Kα radiation is used. The two-dimensional nature of the detector means that it is not necessary, as it is with a four-circle machine, to bring all reflections into a horizontal plane (the ω axis is considered vertical and the incident X-ray beam horizontal), so less movement of the crystal is required to access all reflections. The relative advantages and disadvantages of different goniometer designs, with one, two and three available rotations for the crystal, are hotly debated; certainly, with only one rotation axis, and for some low-symmetry crystal systems and particular crystal orientations, not all reflections may be measurable. There is also no general agreement about the merits of collecting data with rotation about different axes. Not having a full χ-circle does give considerably more freedom for extra facilities such as low-temperature devices.

CCD systems can be combined with different X-ray sources in the same way as four-circle diffractometers: conventional sealed tubes, rotating anodes, new types of micro-focus tubes, and synchrotron radiation are all used.

Corrections need to be applied to raw CCD images for a number of factors. These are usually fully integrated into commercial systems. Some of the corrections are listed here.

(a) Spatial distortion. The demagnification of the diffraction image from the phosphor to the CCD chip via the fibre-optic taper used in most systems is never perfectly linear and to scale. The mapping of CCD pixels to original face-plate positions needs to be calibrated and a correction then applied to every image. The calibration can be made, for example, by recording a pattern of X-rays from an amorphous scatterer or a fluorescent sample through an accurately machined grid of fine holes placed over the detector face. It need be done only once unless the phosphor–taper–CCD assembly is changed in any way.

(b) Non-uniform intensity response. Equal incident X-ray intensity at different points on the detector face may lead to unequal numbers of electrons at the corresponding pixels on the CCD, for various reasons in the different components of the system. Calibration involves recording a uniform intensity 'flood field' and measuring the CCD image for this.

(c) Bad pixels. Minor faults in CCD production can include individual pixels or even rows of pixels which do not respond correctly to incident light. Substantial faults mean an unusable chip, but a few bad pixels can be tolerated and flagged as bad, particularly for systems in which pixels are 'binned' by combining 2×2 or other groups of pixels rather than using all pixels individually.

(d) Dark current. Thermal excitation leads to the generation and trapping of electrons in the CCD pixel wells even when there is no incident light, and this slowly builds up a background 'dark image' on the detector, which will be read out superimposed on the true image. The effect is minimized by cooling the CCD (typically to between $-40°$ and $-80°$) and can be corrected by recording a dark-current image (no X-rays) for the same time as a normal exposure and subtracting this from each measured frame. The dark-current image is temperature dependent and needs to be measured for the appropriate length of time, preferably averaged over several recordings to reduce statistical fluctuations.

6.4   A typical experiment

The steps involved in setting up and collecting data with an area detector are similar to those for a four-circle diffractometer, but there are some important differences, and the whole process is usually much faster.

6.4.1   Crystal screening

Mounting and optically centring a crystal is essentially just as on a four-circle machine. A random-orientation stationary (or small oscillation) exposure can be recorded in a matter of a few seconds and gives an almost instantaneous indication of the quality and intensity of diffraction by the crystal. At this stage, problems such as poor crystallinity, obvious splitting of reflections, and generally weak diffraction can be identified and a different crystal selected. Assessment of the mosaic spread of reflections requires a series of frames to be collected, but this can be achieved within a few minutes, or it can conveniently be done as part of determining the unit cell and crystal orientation in the next step. A substantial queue of poor-quality samples can easily be disposed of in a single morning! It should be noted, however, that CCD systems can often handle samples of poorer quality than four-circle diffractometers and still give an acceptable structural result, because of their greater sensitivity and efficiency overall.

6.4.2   Unit cell and orientation matrix determination

Collection of a series of short exposure frames (preferably covering more than one small region of reciprocal space) takes a few minutes. A typical single frame is shown in Fig. 6.1. The positions of the reflections, including interpolation between successive frames to obtain precise setting angles, are obtained and these are stored in a list by the control program. This is equivalent to the initial reflection search on a four-circle diffractometer and yields coordinates of observed reflections in reciprocal space, referred to goniometer axes. With reasonable-sized structures and moderate intensities, there may be upwards of a hundred reflections, compared with the 10–25 often used on a four-circle.

Indexing of the reflections and determination of the crystal orientation matrix and unit cell parameters follows as described previously, with the advantage of usually having many more reflections available and hence a lower probability of obtaining an incorrect

Fig. 6.1.   A typical short-exposure narrow-angle frame of data from a CCD detector.

cell. The greater number of reflections can also make it easier to obtain a result from twins and other samples which are not single crystals, though the outcome then needs to be examined particularly carefully to see which reflections do not fit the proposed cell and whether there is a clear explanation for this. Assessment of the mosaic spread is usually also part of this step; it is important for decisions of how to collect the full data set and how to extract intensities from the raw data frames.

One very important difference from four-circle machines is that the determination of an orientation matrix in advance of full data collection is not essential, because the complete diffraction pattern is measured rather than just the reciprocal lattice points (Bragg reflections). Obtaining a matrix can be left until the full set of frames is measured, but it must be done before the reflection positions can be predicted and the intensities integrated. An incorrect (or uncertain) initial matrix and cell is not a problem, as their determination can be revised following the full data collection without loss of data.

The unit cell and matrix can be checked just as in four-circle diffractometer experiments, as described in the previous chapter (p. 57).

6.4.3 Data collection

Some parameters need to be set for the full data collection, on the basis of the preliminary screening measurements and unit cell determination. Factors which affect these
decisions include the overall level of intensity (choice of frame measuring times), mosaic
spread (angular width of individual frames, which are measured over a small range of
one goniometer axis rather than stationary), and crystal symmetry (fraction of the whole-
sphere diffraction pattern to be measured). Area detectors offer the advantage of a high
degree of data redundancy, with many symmetry-equivalent reflections and the measurement of the same reflections in different positions; this helps to reduce errors and
also provides information on which to base corrections for systematic effects such as
absorption. Even with a high degree of redundancy, data collections often take only a
few hours.

6.4.4 Data reduction and corrections

Integrated intensities need to be extracted from the raw frames. Integration software
uses the orientation matrix to determine the reflection positions and the estimation of
intensities can exploit the three-dimensional information available for each reflection in
some form of profile fitting. There may also be facilities for updating and refining the
orientation matrix during the integration process, to allow for uncertainties or gradual
changes in the crystal orientation.

As for four-circle diffractometer measurements, corrections need to be made for Lp
factors (which are instrument-specific) and, where appropriate, for absorption, incident beam intensity variations (especially in synchrotron work), and other sources of
systematic errors.

Output of data reduction routines usually includes analysis of data significance
(I/σ ratios as a function of Bragg angle and other variables), and data coverage and
redundancy. The data set is then ready for structure determination.

The high degree of redundancy of data, with individual reflections measured more
than once in different crystal and detector positions and with coverage of symmetry-
equivalent reflections often amounting to the whole sphere of reciprocal space within a
certain maximum Bragg angle, permits a thorough investigation of possible systematic
errors in the intensities and a means of making effective corrections. These are based on
the fact that, in the absence of systematic errors, symmetry-equivalent reflections should
have identical intensities. Errors may be due to a variety of factors, including absorption,
X-ray beam inhomogeneity and intensity fluctuations, synchrotron beam decay, and
instrumental instabilities. They are generally modelled in some empirical way, such as a
sum of spherical harmonics with incident and diffracted beam directions as the function
variables, and scale factors can be calculated and applied to individual recorded frames
to improve the agreement for symmetry equivalents. Improvements can be dramatic, but
the disadvantage is that it is difficult or impossible to unravel the various contributions
to the errors.

7

Fourier syntheses

The calculation of the electron density from the X-ray diffraction pattern of a crystal is accomplished by a Fourier synthesis, which is thus one of the fundamental calculations of X-ray crystallography. It may be regarded as the operation which performs the function of a lens on the scattered X-rays to yield an image of the crystal structure. It exists in a number of different forms, so it is advisable to learn some of its characteristics in order to use it intelligently.

7.1 Fourier synthesis in 1D

For a one-dimensional crystal, this fundamental calculation is described by

$$\rho(x) = \frac{1}{a} \left\{ \sum_h F(h) \exp(-2\pi ihx) \right\} \tag{7.1}$$

where $\rho(x)$ is the density of scattering matter at a point a fraction x from the origin of the unit cell, $F(h)$ is the (complex) structure factor, the summation is over all orders of diffraction, h, and a is the unit cell length. Adding together pairs of terms in the summation with positive and negative h gives

$$F(-h) \exp[2\pi ihx] + F(h) \exp[-2\pi ihx] = |F(h)|\{\exp[2\pi ihx - \phi(h)]$$
$$+ \exp[-2\pi ihx + \phi(h)]\}$$
$$= 2|F(h)| \cos(2\pi hx - \phi(h)). \tag{7.2}$$

The relationships used here are that $|F(h)| = |F(-h)|$, $\phi(h) = -\phi(-h)$, and $\exp(i\phi) = \cos(\phi) + i \sin(\phi)$. This allows the Fourier summation to be expressed as

$$\rho(x) = \frac{1}{a} \left\{ F(0) + 2 \sum_h |F(h)| \cos(2\pi hx - \phi(h)) \right\} \tag{7.3}$$

where the summation is now performed over all positive values of h.

It can be seen that the contribution to the image of each pair of diffracted beams $F(h)$ and $F(-h)$ is a cosine wave. The image is therefore made up of a constant background, given by $F(0)$, modulated by adding waves of the spacings and orientations (in 3D) of the crystal planes corresponding to the diffraction pattern. In fact, the appearance of the diffraction pattern will always give some information about the arrangement of atoms; in particular, those planes which scatter strongly indicate that the arrangement of electrons in the unit cell is something like the arrangement of those planes.

Notice that phases are required to perform the Fourier summation. The crystal symmetry affects the phases in that they can only be 0 or π if the structure is centrosymmetric. In our 1D example for a centrosymmetric structure, the expression for the Fourier synthesis simplifies to:

$$\rho(x) = \frac{1}{a}\left\{ F(0) + 2\sum_h F(h)\cos(2\pi hx) \right\} \tag{7.4}$$

where the phase becomes part of the $F(h)$, so that $F(h)$ is positive for a phase of 0 and negative for a phase of π.

The electron density is normally represented as a discrete function evaluated over a regular grid of points. In order not to lose any detail, the grid over which a Fourier is calculated should have about three points to the resolution of the data. Thus, with data to 1 Å (θ_{max} 22° with Mo Kα radiation), three points per Å are suitable.

7.2 A 1D example—iron pyrites

A simple example of the use of a one-dimensional Fourier is provided by the structure of iron pyrites, FeS_2. It is cubic, $a = 5.40$ Å, space group $Pa\bar{3}$ with four formulae per cell. The Fe atoms must lie on the origin and face-centred positions, while the S atoms lie on threefold axes which run parallel to the body diagonals. The centre of the S–S bond is at $(\frac{1}{2}, \frac{1}{2}, \frac{1}{2})$ and symmetry-related positions, and the S atoms have corresponding coordinates (x, x, x), etc. The positions of the S atoms and hence the length of the S–S bond may be determined by examining the one-dimensional Fourier sum for the reflections $h00$. This gives a centrosymmetric projection onto the x axis in which the symmetry makes the repeat distance half a cell. For the $h00$ data, therefore, we may take a one-dimensional cell with $a = 2.70$ Å and h one-half of the 3D value. There will be one Fe at $x = 0$ and two S atoms per projected cell, at positions to be found.

Since the iron atom position is known, structure factors corresponding to this one atom may be calculated. Their phases (or signs in this case) may then be used as an approximation to the true phases of the complete structure, which allows an approximate Fourier synthesis for iron pyrites to be calculated. Figure 7.1 shows the contribution to the electron density of each term in this Fourier series. Also shown are the magnitudes of the observed structure factors, $|F_o|$, and the structure factors calculated from the iron, F_c, where it can be seen that the signs are all positive.

If the Fourier synthesis is performed using the F_c structure factors, as in Fig. 7.2, only the iron atom should be seen. This is the F_c synthesis. There is certainly a large peak where the iron should be, but the rest of the map does not show a flat background as might be expected, and a lot of the electron density appears to be negative. The ripple is due to series termination, i.e. structure factors for $h = 6, 7, \ldots$ were not included in the summation because their magnitudes were not measured experimentally. In addition, the term for $h = 0$ was not included. This omission makes the mean value in the map equal to zero, giving the appearance of large amounts of negative density.

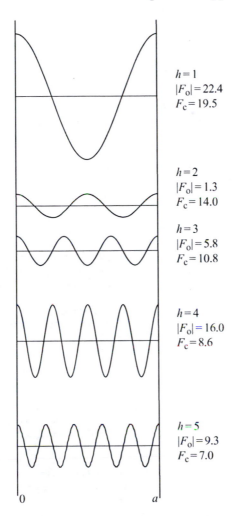

$h=1$
$|F_o|=22.4$
$F_c=19.5$

$h=2$
$|F_o|=1.3$
$F_c=14.0$

$h=3$
$|F_o|=5.8$
$F_c=10.8$

$h=4$
$|F_o|=16.0$
$F_c=8.6$

$h=5$
$|F_o|=9.3$
$F_c=7.0$

0 a

Fig. 7.1.   Representation of $F_o \cos(2\pi h x)$, the contribution of $F(h)$ to $\rho(x)$.

Performing the Fourier synthesis using the observed magnitudes, $|F_o|$, and the calcu-
lated signs (all positive) gives the F_o synthesis, shown in Fig. 7.3. Because the magnitudes
correspond to the complete structure and the signs are largely correct, both the iron and
the sulfur atoms now appear. We hope you have a sense of wonder that atoms can be
observed as closely as this: their size, shape and distance apart can be clearly seen. Note
that the map suffers from series termination in the same way as that in Fig. 7.2.

A third form of Fourier synthesis is shown in Fig. 7.4. It is known as the difference
synthesis because it is calculated using $|F_o| - |F_c|$ as coefficients, together with the cal-
culated signs. It shows the difference between the maps in Figs 7.2 and 7.3. Because
series termination affects the F_o and F_c syntheses in the same way, it cancels out in the
difference synthesis, making it particularly suitable for the identification of unknown
atom positions. The small remaining peak at $x = \frac{1}{2}$ requires an explanation. It appears
because not all the calculated signs are in fact correct. The assumption that the phases

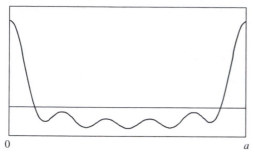

Fig. 7.2. The F_c synthesis, based on the calculated values for F assuming that only the Fe atom is present: $\sum_h F_c \cos(2\pi hx)$. Note that, in addition to the peak corresponding to Fe, there is substantial 'ripple'.

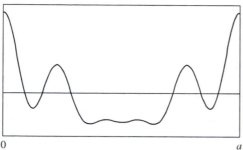

Fig. 7.3. The F_o synthesis, using the observed values for F and the calculated phases (signs are all $+$) from the Fe positions: $\sum_h F_o \cos(2\pi hx)$. The ripple is much the same as in Fig. 7.2, but peaks corresponding to the S positions have appeared.

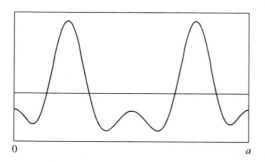

Fig. 7.4. The difference synthesis, the difference between Figs 7.2 and 7.3. The sign of F_o is assumed to be that of F_c and $F_o - F_c$ are used as coefficients: $\sum_h (F_c - F_o) \cos(2\pi hx)$. Note that most of the ripple has disappeared, but that a 'ghost' peak has appeared at $x = \frac{1}{2}$.

of F_o and F_c are the same breaks down more often as the fraction of the structure which is unknown increases, and is discussed below.

This one-dimensional example shows the same features which are important in two- and three-dimensional Fourier sums also.

7.3   The 2D synthesis

In two dimensions, the expression for the Fourier synthesis is:

$$\rho(x, y) = \frac{1}{A} \sum_h \sum_k F(h, k) \exp[-2\pi i(hx + ky)]$$

$$= \frac{1}{A} \sum_h \sum_k |F(h, k)| \cos[2\pi(hx + ky) - \phi(h, k)] \qquad (7.5)$$

where A is the area of the two-dimensional unit cell and h and k are orders of diffraction in the x and y directions respectively. The summations on the right are performed over $h \geq 0$ and all k. It is often possible to make gains in efficiency by using the symmetry of the crystal. In particular, if the plane group (two-dimensional space group) of the structure is centrosymmetric, the summation will simplify as in the one-dimensional case to:

$$\rho(x, y) = \frac{1}{A} \sum_h \sum_k \{F(h, k) \cos[2\pi(hx + ky)]\}. \tag{7.6}$$

Since every structure factor contributes to every point at which the Fourier is evaluated, the amount of calculation becomes rather large in two dimensions; it is considerably more in 3D. This may be reduced by factorizing the summation. To see how this affects the calculation, consider the following alternative forms of the same expression:

$$(a + b)(c + d) = ac + ad + bc + bd. \tag{7.7}$$

The left-hand side requires two additions and one multiplication, while the right requires four multiplications and three additions. Clearly, the factorized form is more efficient.

There are two factorization schemes in common use in crystallographic applications. One is the Beevers–Lipson factorization, which may be written:

$$\rho(x, y) = \frac{1}{A} \left\{ \sum_k \left(\sum_h F(h, k) \exp[-2\pi ihx] \right) \exp[-2\pi iky] \right\} \tag{7.8}$$

where the inner sum is simply the one-dimensional sum seen earlier. This needs to be calculated for all values of x and h. The second stage of the calculation is carried out by performing the sums over k. These are also one-dimensional calculations, thus simplifying the original two-dimensional summation. The other factorization scheme is often called the Fast Fourier Transform (FFT), one version of which is known as the Cooley–Tukey algorithm. This factorization leads to a particularly efficient calculation, and is generally faster than the Beevers–Lipson method. This advantage is somewhat moderated by the facts that the FFT requires all data out to the maximum value of a particular index to be included in the sums, even when the contribution is zero, and that it is much less easily simplified to allow for symmetry in the cell than is the B–L summation. In practice, Fourier summations normally occupy a very small part of crystallographic computing time.

Until about 1960, most crystal structures were solved in projection, as the amount of calculation involved in three-dimensional work was so great for the calculators then available. The three-dimensional structure then had to be inferred from two or three projections. Although this is no longer done in practice, the example given here illustrates simply many of the principles involved in three-dimensional Fourier synthesis. We shall look at the c-axis projection of ammonium oxalate monohydrate, as this axis is sufficiently short that the atoms are reasonably well resolved.

Ammonium oxalate monohydrate, $(NH_4)_2C_2O_4 \cdot H_2O$, gives orthorhombic crystals, space group $P2_12_12$, with $a = 8.03$, $b = 10.31$, $c = 3.80\,\text{Å}$, $Z = 2$. The structure has frequently been examined (see especially J. H. Robertson, *Acta Cryst.*, 1965, **18**, 410). The symmetry of the c-axis projection is the centrosymmetric plane group

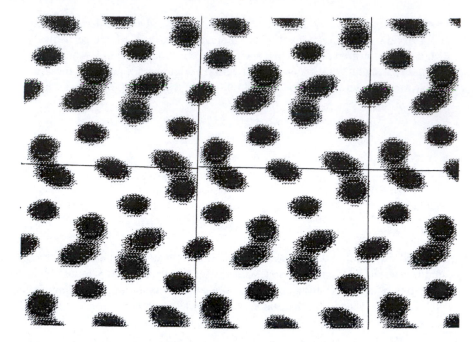

Fig. 7.5. Fourier map based on 19 data to a resolution of 2 Å for ammonium oxalate monohydrate (c axis (3.8 Å) projection: $a = 7.98$, $b = 10.04$ Å, plane group $p2gg$).

$p2gg$ (a primitive rectangular cell with glide lines perpendicular to both axes; for further details see Exercise 2.5 at the end of Chapter 2). In all cases of Fourier maps shown here, the data are phased correctly, and an arbitrary grey scale has been used to indicate electron density.

Figure 7.5 is a low-resolution (2 Å) electron density map calculated from only 19 structure factors. This is below atomic resolution and the carbon atoms cannot be seen as separate peaks from the oxygens. This resolution is typical of that available in protein crystallography. The structure is more or less recognizable but, as will be seen later, it is difficult to solve structures when data are only available to this resolution.

Figure 7.6 shows what is normally available from a good structure determination. The resolution is now 0.8 Å, using 120 structure factors, and all the non-hydrogen atoms are clearly visible as separate peaks. The effect of higher resolution data is to locate the non-hydrogen atoms well, although determining their positions from an electron density map is only the first step in creating an atomic model of the structure. The atomic coordinates so obtained are normally subjected to least squares refinement, which results in much more accurate values. The hydrogen atom positions are often better defined in a map with a slightly lower resolution, as we shall see, although they can in fact be detected in the map given!

The final map, shown in Fig. 7.7, is an E-map with phases determined by direct methods. The structure factors have been modified to make the intensities independent of θ, thus generating E values, so high-angle data have a proportionately large effect.

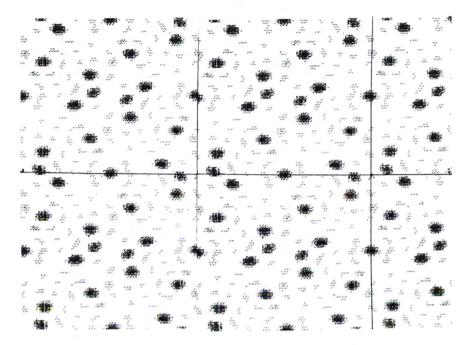

Fig. 7.6.   Map based on 120 data to a resolution of 0.8 Å.

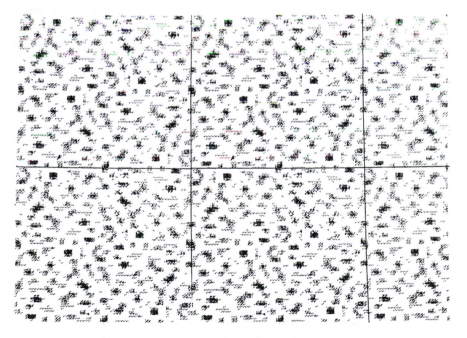

Fig. 7.7.   Map based on 29 high E data to a resolution of 0.6 Å.

In addition, the phases of only the strongest structure factors (largest E values) are normally determined, so this map is calculated leaving out a large number of the weaker reflections. The structure has clearly been solved, although the noise level is high. The strongest peaks all correspond to atoms, but some spurious peaks are very obvious, especially the 'bridge' between the oxygen atoms of the oxalate. The water molecule is the lowest genuine peak.

7.4 The 3D synthesis

The general expression for a three-dimensional Fourier synthesis is:

$$\rho(x, y, z) = \frac{1}{V} \sum_h \sum_k \sum_l \{F(h, k, l) \exp[-2\pi i(hx + ky + lz)]\}$$

$$= \frac{1}{V} \sum_h \sum_k \sum_l \{|F(h, k, l)| \cos[2\pi(hx + ky + lz) - \phi(h, k, l)]\} \quad (7.9)$$

where V is the volume of the unit cell, and $\rho(x, y, z)$ is the relative electron density at the point (x, y, z). The summations in the second expression are carried out over all k and l with $h \geq 0$.

A three-dimensional Fourier summation involves a much lengthier calculation than do one- and two-dimensional calculations, so an efficient algorithm is essential. Both the Beevers–Lipson and the FFT are used, but the FFT gains in speed over the Beevers–Lipson as the size of the map increases. It is, therefore, the method of choice for large electron density maps.

The interpretation of a Fourier is usually done by computer searching of computer-generated maps to locate peaks. In well-behaved cases, this is adequate to give good approximations to the centres of atom positions. Otherwise, with disordered structures or only low-resolution data, it is often misleading, because the extent and shape of the peaks are not normally evaluated, only the positions of their maxima. An example is Fig. 7.5, where a peak search would find only four atoms in the oxalate ion!

7.5 Uses of Fouriers

The major problem with Fourier syntheses is always that the phases, which are vital to the summations, are only known approximately at best, while the exact values of the amplitudes make relatively little difference. Since a Fourier synthesis is generally the only way of knowing whether a structure has been solved, most of the variations in uses are attempts to maximize useful data and minimize errors. The main uses of Fourier syntheses in crystal structure determination are to calculate the following.

1. Patterson functions. These are discussed in the next chapter.
2. Initial maps for a few data phased by direct methods. With a map phased by direct methods, the main choice is whether to use Es or Fs as amplitudes (see Chapter 9 on direct methods), together with the determined phases. It is usual to use Es, principally because these are more readily available at this stage of the calculations, and they give

sharper peaks. The slight disadvantage is that they also give a noisier background to
the map.

3. Difference maps to expand partially solved structures. Difference maps for partial
   structures, as was used in the iron pyrites example, need attention when the known
   part of the structure is relatively small. Care must be taken that the calculated and
   observed structure factors are on the same scale. In addition, intensities for which
   $I < 2\sigma(I)$ should probably not be used at all, as the error in the measurement will
   be greatly magnified in the difference.

4. Location of hydrogen atoms. When hydrogen atoms are being sought, the structure is
   almost completely known. Difficulties arise because the scattering from the hydrogen
   atoms is so small that it will often be lost in the noise. This is particularly true for non-
   centrosymmetric structures, where the absence of the hydrogen atoms in a model is
   partially compensated by shifts in the phases. High-resolution data are not useful for
   the location of hydrogen atoms, since the hydrogen atom contribution drops rapidly
   as $\sin\theta$ increases. It is often useful to eliminate (or down-weight) the high-angle data;
   generally retaining only data to a resolution of about 1.5 Å will be satisfactory.

5. Final checking of a structure. At the end of a crystal structure determination, a
   difference Fourier must be calculated to check for unexplained electron density in the
   map. Clearly, this must be an unweighted map and must contain all data which have
   been used in the structure refinement. The commonest cause for residual density
   is failure to correct the data properly for absorption. Otherwise, residual density
   probably does indicate a poor model for the structure, particularly badly modelled
   disorder or possibly twinning.

7.6   Weighted Fouriers

When only part of a structure is known, the standard method of determining the rest is to
calculate an electron density map in which some (or all) of the unknown atoms appear.
We will look at how this information on unknown atomic positions is obtained.

Let FN be a structure factor of the complete structure, FP the contribution to FN of the
known atoms and FQ the contribution of the unknown atoms. Normally, FP(F_c) will be
known in amplitude and phase, but only the amplitude of FN($|F_o|$) will be known. It is
required to estimate FQ to obtain information about the unknown atoms. These quantities
are related by FN = FP + FQ as shown in Fig. 7.8 for a centrosymmetric structure.

There are two possibilities: FN and FP have either the same sign or different signs. The
diagram shows that when the signs are the same, FQ has a smaller magnitude than when
they are different. It is known from structure factor statistics that the smaller magnitude

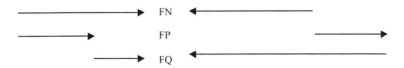

Fig. 7.8.   The two possibilities of FN and FP having the same or different signs.

is the more likely, so the signs are more likely to be the same. The uncertainty in this estimate of sign will depend upon the number of unknown atoms and the magnitudes of FN and FP. A proper way of taking this into account is to calculate a weight w for each structure factor and to use this in the calculation of a weighted Fourier synthesis. If $X = |FN \cdot FP| / \sum f^2$ where the summation is over the missing atoms, then

$$w = \tanh(X) \qquad (7.10)$$

and the coefficients of the Fourier are $w|FN|s(FP)$, where $s(FP)$ is the sign of FP. This should show all the known atoms and a number of the unknown atoms depending upon how good the estimates of the signs are.

In a non-centrosymmetric space group, the structure factors FN, FP and FQ are related as in Fig. 7.9. It is seen that FQ depends upon the unknown angle α, the difference between the phases of FN and FP. Averaging over all possible values of α gives the best estimate of FQ as:

$$FQ = 2(w|FN| - |FP|)\exp(iFP). \qquad (7.11)$$

The weight w depends upon the quantity X defined in equation (7.10) for centrosymmetric structures and is shown in Fig. 7.10. It is clear that the fewer the missing atoms or the stronger the magnitudes $|FN|$ or $|FP|$, the better is the estimate of phase.

A weighted Fourier map calculated using coefficients obtained from (7.11) should show the locations of the unknown atoms contributing to FQ. This is usually best for a

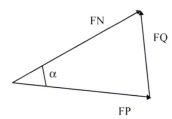

Fig. 7.9.   The structure factors FN, FP and FQ for a non-centrosymmetric structure.

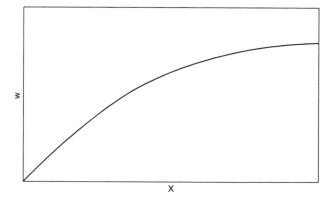

Fig. 7.10.   Weighting function for Fourier syntheses.

structure in which the known atoms are much heavier than the others. When the known and unknown parts of the structure consist of similar atoms, better results are often obtained for a centrosymmetric structure by calculating a map using the coefficients given in (7.10). This should show both known and unknown atoms on a similar scale. For a non-centrosymmetric structure, a rearrangement of (7.11) shows that $(2w|FN| - |FP|)\exp(iFP) = FP + FQ$. Using these as coefficients gives a so-called '$2F_o - F_c$' map, which clearly contains information about the unknown atoms as well as peaks corresponding to the atoms used in the calculation of the phases.

Exercises

7.1 Factorize the expressions $\cos 2\pi(hx + ky + lz)$ and $\sin 2\pi(hx + ky + lz)$ into terms involving products of one-dimensional terms. Consider the space group $Pmmm$, whose general equivalent positions are given below. How will the expression be simplified when symmetry is taken into consideration?

$$x, y, z \qquad -x, -y, z \qquad -x, y, -z \qquad x, -y, -z$$
$$-x, -y, -z \qquad x, y, -z \qquad x, -y, z \qquad -x, y, z$$

7.2 Use the one-dimensional Fourier map for FeS_2 to determine the S–S bond length and the minimum $Fe \cdots S$ distance in the crystal. Note that the projection results in an ambiguity and that a coordinate x in one dimension implies x, x, x in three dimensions.

7.3 The undertaking of a Fourier synthesis in three dimensions is too lengthy a process for hand calculation. Some feel for it, however, can be obtained by attempting a one-dimensional synthesis. For a centrosymmetric structure, the summation reduces to

$$\rho(x) = \sum_h F_h \cos 2\pi hx \qquad (7.12)$$

where $\rho(x)$ is the electron density (unscaled) at a point x along the line, F_h is the (signed) structure factor for the reflection of order h, and the sum is carried out over as many h as possible. This may be accomplished easily by using 'strips' of the Beevers–Lipson type, where each strip contains values of $F_h \cos 2\pi hx$ at a range of values of x for particular values of F and h. These strips are aligned to enable the sum to be done for a particular value of x, the result being the one-dimensional electron density at that point. Strips of a similar sort (containing values of $f_j \cos 2\pi hx_j$ at a range of values of h for particular values of f and x) could be used for a structure factor calculation, giving the contributions to

$$F_h = \sum_j f_j \cos 2\pi hx_j \qquad (7.13)$$

where f_j is the scattering factor for the jth atom having a coordinate x_j. In this case, j strips will be needed to complete the sum for each value of h.

An example of a structure where such a summation may usefully be carried out is DL-3-bromooctadecanoic acid (S. Abrahamsson and M. M. Harding, Acta Cryst., 1966, **20**, 377). It is triclinic, space group $P\bar{1}$, with $a = 5.68, b = 5.63, c = 32.80$ Å, $\alpha = 101.8°$, $\beta = 93.1°, \gamma = 97.9°$. The projection of the structure on $0, 0, z$ gives an interpretable map of the two molecules in the unit cell, related by centres of symmetry at $0, 0, 0$ and $0, 0, \frac{1}{2}$. 'Strips' of data relating to the strongest $F(00l)$ data are given below as horizontal lines of numbers (Table 7.1). They are arranged so that the sum of all the numbers on one line will give the value of the electron density for that value of z.

Fourier syntheses

Table 7.1 Values of $|F|\cos 2\pi lz$

z	l									
	4	5	6	9	10	11	14	15	20	21
0.00	24	18	23	16	17	15	10	13	13	26
0.01	24	18	22	13	13	12	6	7	4	6
0.02	21	15	17	7	5	3	−2	−4	−10	−22
0.03	18	11	10	−2	−5	−7	−9	−12	−10	−18
0.04	13	6	2	−10	−13	−14	−10	−10	4	14
0.05	8	0	−7	−15	−17	−14	−3	0	13	24
0.06	2	−6	−15	−15	−13	−8	6	10	4	−2
0.07	−4	−11	−21	−11	−5	2	10	12	−10	−25
0.08	−10	−15	−23	−3	5	11	7	4	−10	−11
0.09	−16	−18	−23	6	13	15	−1	−7	4	20
0.10	−20	−18	−19	13	17	12	−8	−13	13	20
0.11	−23	−18	−12	16	13	4	−10	−7	4	−9
0.12	−24	−15	−4	14	5	−6	−4	4	−10	−25
0.13	−24	−11	4	8	−5	−14	4	12	−10	−3
0.14	−23	−6	12	−1	−13	−14	10	10	4	24
0.15	−20	0	19	−9	−17	−9	8	0	13	15
0.16	−16	6	23	−15	−13	1	1	−10	4	−16
0.17	−10	11	23	−16	−5	10	−7	−12	−10	−23
0.18	−4	15	21	−12	5	15	−10	−4	−10	5
0.19	2	18	15	−4	13	13	−6	7	4	26
0.20	8	18	7	5	17	5	3	13	13	8
0.21	13	18	−2	12	13	−5	10	7	4	−22
0.22	18	15	−10	16	5	−13	9	−4	−10	−19
0.23	21	11	−17	14	−5	−15	2	−12	−10	12
0.24	24	6	−22	8	−13	−10	−6	−10	4	25
0.25	24	0	−23	0	−17	0	−10	0	13	0
0.26	24	−6	−22	−8	−13	10	−6	10	4	−25
0.27	21	−11	−17	−14	−5	154	2	12	−10	−12
0.28	18	−15	−10	−16	5	13	9	4	−10	19
0.29	13	−18	−2	−12	13	6	10	−7	4	22
0.30	8	−18	7	−5	17	−5	3	−11	1	−8
0.31	2	−18	15	4	13	−13	−6	−7	4	−26
0.32	−4	−15	21	12	5	−15	−10	4	−10	−5
0.33	−10	−11	23	16	−5	−10	−7	12	−10	23
0.34	−16	−6	23	15	−13	−1	1	10	4	16
0.35	−20	0	19	9	−17	9	8	0	13	−15
0.36	−23	6	12	1	−13	14	10	−10	4	−24
0.37	−24	11	4	8	−5	14	4	−12	−10	3
0.38	−24	15	−4	−14	5	6	−4	−4	−10	25
0.39	−23	18	−12	−16	13	−4	−10	7	4	9
0.40	−20	18	−19	−13	17	−12	−8	13	13	−21
0.41	−16	18	−23	−6	13	−15	−1	7	4	−20
0.42	−10	15	−23	3	5	−11	7	−4	−10	11
0.43	−4	11	−21	11	−5	−2	10	−12	−10	25

Table 7.1 Continued

z										
0.44	2	6	−15	15	−13	8	6	−10	4	2
0.45	8	0	−7	15	−17	14	−3	0	13	−24
0.46	13	−6	2	10	−13	14	−10	10	4	−14
0.47	18	−11	10	2	−5	7	−9	12	−10	18
0.48	21	−15	17	−7	5	−3	−2	4	−10	22
0.49	24	−18	22	−13	13	−12	6	−7	4	−6
0.50	24	−18	23	−16	17	−15	13	−13	13	−26

(a) Evaluate the Patterson function (using $|F|$ in place of F^2), by taking all signs as given (zero phase for all reflections). This will give a large peak at the origin and one other outstanding peak in the asymmetric unit. Find the coordinate of the bromine atom by taking the coordinate of this peak as $2z$ for bromine (see the next chapter for further details of the Patterson function).

(b) Use the data to evaluate the signs of $F(00l)$ based on this bromine position. They will be those on the strip corresponding most closely to the bromine z coordinate (why?).

(c) Apply these signs (i.e. invert the signs in each column that corresponds to a negative value of F). Now calculate the sums again with these signs; this is the F_o synthesis based on phases provided by the bromine atom alone. Sketch the answers as a function of z and suggest an interpretation of the pattern. In the actual structure, the chains of the molecules are fully extended, and are inclined at an angle of about $20°$ to the c axis.

8

Structure determination by Patterson methods

The 'solution' of a crystal structure is almost always a result of successive approximations, making use of the fact that the repeating structure and the diffraction pattern are
related to one another by Fourier summations. A structure is generally said to be solved
when most of the atoms have been located in the unit cell. In general, at the outset the
usual situation is:

structure	diffraction pattern
known cell contents	*known amplitudes*
unknown positions	*unknown phases*

The knowledge of a few atom positions will enable better phases to be calculated for the
whole diffraction pattern, resulting in a better electron density map, and revealing further
atoms. Similarly, approximate values for some phases will make it possible to calculate
a map in which the positions of at least some atoms may be determined. The two most
important methods of structure determination, 'Patterson' and 'direct' methods, differ
in that the first starts by obtaining some information about atom positions, while the
second starts from some information about phases.

The basis of the Patterson methods is very simple. It is that the phase problem is
simply ignored. As discussed in the last chapter, the electron density at any point in the
unit cell (ρ) may be calculated as:

$$\rho(xyz) = (1/V) \sum_h \sum_k \sum_l \{|F(hkl)| \cos[2\pi(hx + ky + lz) - \phi(hkl)]\} \qquad (8.1)$$

where V is the unit cell volume, and the summation is done over all data $F(hkl)$, $\phi(hkl)$.
The phases are not, of course, known at the beginning, but a summation which can
always be carried out is:

$$P(uvw) = (1/V) \sum_h \sum_k \sum_l \{|F(hkl)|^2 \cos[2\pi(hu+kv+lw)]\}. \qquad (8.2)$$

This function is called the 'Patterson function' after its inventor, A. L. Patterson. Its
meaning may be visualized by remembering that the amplitude of each individual structure factor contains the information as to how efficiently one particular set of planes
acts as a diffraction grating for X-rays. This means that large amplitudes correspond to
planes whose spacings and orientations are strongly reflected in the atomic arrangement.
The phases are only needed in order to combine that information into a recognizable

representation of the structure. If all the phases are set to zero (or to any other value), the information will remain but will not be so easy to interpret.

What this particular function represents may be seen in one dimension as follows. Consider the structure factor equations for $F(h)$ and $F(-h)$.

$$F(h) = \sum_m f_m \exp(2\pi ihx_m) \qquad F(-h) = \sum_n f_n \exp(-2\pi ihx_n)$$

$$|F(h)|^2 = F(h) \times F(-h) = \sum_m \sum_n f_m f_n \exp(2\pi ih(x_m - x_n)) \qquad (8.3)$$

Or, in words, the set of values $|F(h)|^2$ comprises the structure factors for a structure consisting of objects whose magnitudes are the products of those of all pairs of atoms in the structure $f_m f_n$, and whose positions are the vectors between those atom positions, $(x_m - x_n)$. The same argument may be carried out in three dimensions.

The use of *uvw* in place of *xyz* is to emphasize that the 'space' in which the Patterson summation is carried out is not ordinary space. The result of Fourier transforming this function then is that peaks in the map, which has the same unit cell dimensions as an ordinary Fourier map, no longer represent electron density *positions*, but electron density *vectors*. In other words, a high value represents a high probability of electrons being separated in the structure by the same vector as separates the peak from the origin of the unit cell, and on an absolute scale, densities are normally measured in $e^2 \text{ Å}^{-3}$.

The build-up of a Patterson map may be visualized alternatively as moving each interatomic vector in turn to the origin of the unit cell without changing its direction. There are several features of the resulting map.

1. For n atoms in the unit cell, there will be n^2 possible peaks in the vector map.
2. Each peak will have a height proportional to the product of the atomic numbers of the atoms whose separation it represents.
3. Normally, not all of these will be fully resolved, and in any case, n of them will coincide at the origin, where the highest peak in the map will always occur.
4. Since all phases are zero, or since every vector A→B is matched by a vector B→A, the map will always have an inversion centre at the origin.
5. Screw axes and glide planes in the structure become rotation axes and mirror planes in the Patterson.

Of these, points 4 and 5 affect the possible space groups for a Patterson function. Since $\phi = 0$ for all data, a Patterson space group must be centrosymmetric. Translational symmetry also disappears, since a 2_1 axis implies that $\phi(hkl) = \pi + \phi(h\bar{k}l)$. The result of these two conditions is that the only space groups which can represent Patterson functions are those 24 given in Table 8.1.

Despite the fact that the Patterson function often does not retain the space group of the structure, it can frequently give important information in cases in which the space group is ambiguous. Some of these will be mentioned below.

Clearly, a vector map becomes difficult to interpret unambiguously as the number of atoms increases, but there are two situations in which a Patterson map is especially useful.

Table 8.1　Patterson space groups

Actual space groups	Patterson group (with corresponding space group number)
$P1$ or $P\bar{1}$	$P\bar{1}$ (2)
All monoclinic P	$P2/m$ (10)
All monoclinic C	$C2/m$ (12)
All orthorhombic P	$Pmmm$ (47)
All orthorhombic C	$Cmmm$ (65)
All orthorhombic F	$Fmmm$ (69)
All orthorhombic I	$Immm$ (71)
Tetragonal P, class 4, $\bar{4}$ or $4/m$	$P4/m$ (83)
Tetragonal I, class 4, $\bar{4}$ or $4/m$	$I4/m$ (87)
Tetragonal P, class 422, $4mm$, $\bar{4}m2$ or $4/mmm$	$P4/mmm$ (123)
Tetragonal I, class 422, $4mm$, $\bar{4}m2$ or $4/mmm$	$I4/mmm$ (139)
Trigonal P, class 3 or $\bar{3}$	$P\bar{3}$ (147)
Trigonal R, class 3 or $\bar{3}$	$R\bar{3}$ (148)
Trigonal P, class 312, $31m$ or $\bar{3}1m$	$P\bar{3}1m$ (162)
Trigonal P, class 321, $3m1$ or $\bar{3}m1$	$P\bar{3}m1$ (164)
Trigonal R, class 32, $3m$ or $\bar{3}m$	$R\bar{3}m$ (166)
Hexagonal, class 6, $\bar{6}$ or $6/m$	$P6/m$ (175)
Hexagonal, class 622, $6mm$, $\bar{6}m2$ or $6/mmm$	$P6/mmm$ (191)
Cubic P, class 23 or $m\bar{3}$	$Pm\bar{3}$ (200)
Cubic F, class 23 or $m\bar{3}$	$Fm\bar{3}$ (202)
Cubic I, class 23 or $m\bar{3}$	$Im\bar{3}$ (204)
Cubic P, class 432, $\bar{4}3m$ or $m\bar{3}m$	$Pm\bar{3}m$ (221)
Cubic F, class 432, $\bar{4}3m$ or $m\bar{3}m$	$Fm\bar{3}m$ (225)
Cubic I, class 432, $\bar{4}3m$ or $m\bar{3}m$	$Im\bar{3}m$ (229)

1. When a small number of atoms in the unit cell are significantly heavier than the others, and the structure is, to a first approximation, the structure of those atoms alone. This is called the *heavy atom method*.
2. When a substantial group of atoms have a known arrangement such that their intramolecular vectors may easily be calculated, and only their orientation is unknown. This is called the *Patterson search method*.

An invariant feature of a Patterson map is the large peak at the origin. Of the n^2 peaks in the cell, n will be coincident at the origin, which represents the separation of each atom from itself. Its height will be proportional to $\sum Z^2$ for all the atoms in the unit cell, so it can be useful to get a rough idea of the scale of the Patterson peak heights. For most purposes this is not a serious problem. There are various ways to remove it. One of these is to determine the mean value of $|F^2|$ as a function of $\sin\theta$, $\langle|F^2|\rangle_\theta$, and to use $[|F^2| - \langle|F^2|\rangle_\theta]$ as coefficients rather than $|F^2|$, noting that some of the coefficients will now have a negative sign.

This alteration can readily be combined with another modification of the Patterson function, 'sharpening' it. A 'fully sharpened' Patterson is one in which the normalized structure factors $|E^2|$ are used in place of $|F^2|$. This has the advantage of giving much greater weight to the higher resolution data, and resolving some peaks in the vector map that would otherwise be continuous. It has the disadvantage that, as the high resolution data are also less accurately known, and as they are adjacent to data which have not been measured, there will usually be some spurious definition in the map. A frequently adopted compromise is to use $|EF|$ as Patterson coefficients. If scaling has been done correctly, a sharpened, origin-removed Patterson may be calculated using $|E^2|-1$ as coefficients.

8.1   The heavy atom method

The simplest example of the application of the Patterson function is that of a molecule with a single heavy atom, crystallizing in space group $P\bar{1}$ with two molecules in the unit cell, related by the inversion centre. If the coordinates of one of these atoms is x, y, z (and those of the other $-x$, $-y$, $-z$), there should be, in addition to the origin peak, two equal, outstanding peaks in the map at $(u, v, w) = (2x, 2y, 2z)$ and $(-2x, -2y, -2z)$. Notice that the peak $2x, 2y, 2z$ does not define a unique position, since u is indistinguishable from $1 + u$. There is thus a choice of eight positions with the same value of $2x, 2y, 2z$:
$x, y, z; x, y, \frac{1}{2}+z; x, \frac{1}{2}+y, z; x, \frac{1}{2}+y, \frac{1}{2}+z; \frac{1}{2}+x, y, z; \frac{1}{2}+x, y, \frac{1}{2}+z; \frac{1}{2}+x, \frac{1}{2}+y, z;$
and $\frac{1}{2} + x, \frac{1}{2} + y, \frac{1}{2} + z$. In practice, with a single pair of atoms to be determined, all of these are equivalent, equally correct solutions, as they all represent different choices of the inversion centre that defines the origin of the unit cell. In addition, as any atom at x, y, z implies an identical one at $-x$, $-y$, $-z$, any of these positions may be used as the chosen position for the atom. In general, there are thus 16 positions in the unit cell which may arbitrarily be chosen for the location of the atom found by interpreting the Patterson function!

A more interesting problem arises when there are two independent heavy atoms in $P\bar{1}$ with coordinates x, y, z (A) and x', y', z' (B), along with $-x, -y, -z$ (C) and $-x', -y', -z'$ (D). Letting $\sigma x = x + x'$ and $\delta x = x - x'$, the sixteen vectors from the four atoms will then be:

A–A: $0, 0, 0$	A–B: $\delta x, \delta y, \delta z$
B–A: $-\delta x, -\delta y, -\delta z$	B–B: $0, 0, 0$
C–A: $-2x, -2y, -2z$	C–B: $-\sigma x, -\sigma y, -\sigma z$
D–A: $-\sigma x, -\sigma y, -\sigma z$	D–B: $-2x', -2y', -2z'$
A–C: $2x, 2y, 2z$	A–D: $\sigma x, \sigma y, \sigma z$
B–C: $\sigma x, \sigma y, \sigma z$	B–D: $2x', 2y', 2z'$
C–C: $0, 0, 0$	C–D: $-\delta x, -\delta y, -\delta z$
D–C: $\delta x, \delta y, \delta z$	D–D: $0, 0, 0$

Inspection will show that these reduce to a peak of quadruple weight at $(0, 0, 0)$, pairs of peaks of double weight at $\pm(\sigma x, \sigma y, \sigma z)$ and $\pm(\delta x, \delta y, \delta z)$, and pairs of single weight peaks at $\pm(2x, 2y, 2z)$ and $\pm(2x', 2y', 2z')$. These are shown diagrammatically in Fig. 8.1.

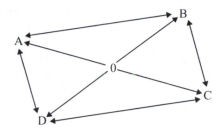

Fig. 8.1. Vectors for two independent heavy atoms in space group $P\bar{1}$.

As above, n of the n^2 vectors are coincident with the origin. In general, symmetry considerations will limit the number of peaks further. Since this structure is centrosymmetric, there will be n peaks of single weight in positions $2x, 2y, 2z$, etc., and the remaining $n^2 - 2n$ peaks will in fact consist of $\frac{1}{2}(n^2 - 2n)$ peaks of double weight at positions $x - x', y - y', z - z'$, etc., between atoms not related to one another by the inversion centre. In this case, one of the atom positions may be chosen to lie in any of 16 positions as above. The other atom will now be restricted to one of two positions related as x', y', z' and $-x', -y', -z'$.

A fairly common example in coordination chemistry consists of one molecule in the asymmetric unit of an orthorhombic cell. The most common of these space groups is $P2_12_12_1$ (non-centrosymmetric). The general equivalent positions are:

$$A: x, y, z \qquad\qquad B: \tfrac{1}{2} + x, \tfrac{1}{2} - y, -z$$
$$C: \tfrac{1}{2} - x, -y, \tfrac{1}{2} + z \qquad D: -x, \tfrac{1}{2} + y, \tfrac{1}{2} - z$$

With four atoms to be considered, there are 16 vectors, of which four coincide at the origin. The remaining 12 will be:

A–B: $\tfrac{1}{2}, \tfrac{1}{2} + 2y, 2z$ A–D: $2x, \tfrac{1}{2}, \tfrac{1}{2} + 2z$ A–C: $\tfrac{1}{2} + 2x, 2y, \tfrac{1}{2}$

B–A: $\tfrac{1}{2}, \tfrac{1}{2} - 2y, -2z$ D–A: $-2x, \tfrac{1}{2}, \tfrac{1}{2} - 2z$ C–A: $\tfrac{1}{2} - 2x, -2y, \tfrac{1}{2}$

C–D: $\tfrac{1}{2}, \tfrac{1}{2} - 2y, 2z$ C–B: $-2x, \tfrac{1}{2}, \tfrac{1}{2} + 2z$ D–B: $\tfrac{1}{2} - 2x, 2y, \tfrac{1}{2}$

D–C: $\tfrac{1}{2}, \tfrac{1}{2} + 2y, -2z$ B–C: $2x, \tfrac{1}{2}, \tfrac{1}{2} - 2z$ B–D: $\tfrac{1}{2} + 2x, -2y, \tfrac{1}{2}$

It should be noted that these peaks have mmm (D_{2h}) symmetry, i.e. the Patterson is centrosymmetric and free of translation symmetry even when the structure is not, and there are thus only three independent peaks to locate. In the list above, each column consists of a peak related to three others by a twofold rotation axis about x, y and z respectively. Also, note that all of these peaks are restricted to special planes—in this case, where u, v or w is $\frac{1}{2}$. Peaks of this type are called Harker peaks, and ones of this type (Harker planes) occur when two atoms are related by a rotation or screw axis. It is thus, in principle, easy to find them, and to check that any trial set of values x, y, z gives all of its predicted peaks.

As a final example, consider a single heavy atom in the centrosymmetric space group *Pbca*. The equivalent positions are:

A: x, y, z 　　　　　　　　　B: $\frac{1}{2} + x, \frac{1}{2} - y, -z$

C: $\frac{1}{2} - x, -y, \frac{1}{2} + z$ 　　　D: $-x, \frac{1}{2} + y, \frac{1}{2} - z$

E: $-x, -y, -z$ 　　　　　　F: $\frac{1}{2} - x, \frac{1}{2} + y, z$

G: $\frac{1}{2} + x, y, \frac{1}{2} - z$ 　　　H: $x, \frac{1}{2} - y, \frac{1}{2} + z$

i.e. those for $P2_12_12_1$ plus an inversion centre. The somewhat imposing total of 64 Patterson peaks is, in fact, not so daunting. Eight will now be coincident with the origin, and the 12 single-weight peaks found for $P2_12_12_1$ will now be double-weight, since, for example, A–B is exactly equivalent to F–E. There are still 32 peaks. Eight of these will be single weight, corresponding to atoms related by an inversion centre:

A–E: $2x, 2y, 2z$ 　　　　　　　B–D: $2x, -2y, -2z$

E–A: $-2x, -2y, -2z$ 　　　　　D–B: $-2x, 2y, 2z$

C–G: $-2x, -2y, 2z$ 　　　　　D–H: $-2x, 2y, -2z$

G–C: $2x, 2y, -2z$ 　　　　　　H–D: $2x, -2y, 2z$

The remaining 24 peaks form another set of Harker peaks, this time six peaks of quadruple weight, characteristic of atoms related by glide or mirror planes. The peaks lie on special lines (Harker lines) and have only one variable:

$$\text{A–F, B–E, G–D, H–C: } \tfrac{1}{2} + 2x, \tfrac{1}{2}, 0$$

$$\text{A–H, D–E, G–B, F–C: } 0, \tfrac{1}{2} + 2y, \tfrac{1}{2}$$

$$\text{A–G, C–E, H–B, F–D: } \tfrac{1}{2}, 0, \tfrac{1}{2} + 2z$$

$$\text{F–A, E–B, D–G, C–H: } \tfrac{1}{2} - 2x, \tfrac{1}{2}, 0$$

$$\text{H–A, E–D, B–G, C–F: } 0, \tfrac{1}{2} - 2y, \tfrac{1}{2}$$

$$\text{G–A, E–C, B–H, D–F: } \tfrac{1}{2}, 0, \tfrac{1}{2} - 2z$$

In summary, for a single heavy atom in a general position in *Pbca* (or any primitive, centrosymmetric, orthorhombic space group) there are 7 independent peaks:

　　1 peak of weight 8 at the origin
　　6 peaks of weight 4 on Harker lines, symmetry related in pairs (3)
　　12 peaks of weight 2 on Harker planes, symmetry related in fours (3)
　　8 peaks of weight 1 in general positions, all symmetry related (1)

A problem occasionally arises when an atom is near a position which makes general and Harker peaks that are indistinguishable. Consider a structure in *Pbca* with a heavy atom at $x = 0.1, y = 0.3, z = 0.25$. There will be peaks of type A–E at 0.2, 0.6, 0.5 and of type A–C at 0.7, 0.6, 0.5. But which is which? Interpreting these peaks the other way round gives equally plausibly $x = 0.35, y = 0.3, z = 0.25$, which is not equivalent to the correct answer. In fact, the vector set of the heavy atoms alone is equivalent, and there is an initially hopeful solution which will not, however, develop other atoms. Be particularly wary of solutions with x, y or z near 0, $\frac{1}{4}$ or $\frac{1}{2}$!

The heavy atom method works best when the peaks in the Patterson depending only on the heavy atoms are readily identified. This will usually happen if the asymmetric unit contains one atom such as copper ($Z = 29$) or a heavier atom, in the presence of 20–30 atoms not heavier than oxygen ($Z = 8$). This is because the ratio of the height of a Cu–Cu vector peak to that of a Cu–O vector peak is $(29 \times 29)/(29 \times 8) = 3.6$. Problems may arise if four or more Cu–O vectors happen to be parallel, and this is by no means uncommon with coordination compounds. Sometimes, several attempts must be made to interpret the vector pattern. If the 'heavy atom' is much lighter, say P or S ($Z = 15$ or 16), the correct vectors may sometimes be identified, but the phases calculated based on the trial positions are not usually good enough to give an interpretable Fourier map. A different problem arises when the heavy atoms are on special positions so that they do not contribute to all data. In all these cases, progress can often be made by using the approximately determined phases along with a direct methods approach for other phases, to obtain an interpretable map.

8.2 Patterson search techniques

The other main application of Patterson methods is in 'Patterson search' methods, when the arrangement of atoms in part of a structure is known from previous work. Such a fragment will have a characteristic vector pattern which can often be a powerful tool in structure solving. A molecule, a group or a ligand will often be suitable, particularly if it has a fused, non-planar ring system. Consider, for example, the camphor molecule (Fig. 8.2).

This molecule is chiral, but has symmetry near mm (C_{2v}). The intramolecular vectors, especially those with a length of 5Å or less, will be complex, and will overlap significantly, but they can vary very little in relative orientation from one structure to another. It is thus usually possible to search a Patterson map automatically and solve quite complex structures with the knowledge that such a group exists in the structure being solved. These methods have been widely developed in macromolecular crystallography. There are two search stages.

8.2.1   Rotation search

Initially, a calculated vector map is made from the coordinates of the 'known fragment'. In this stage, only the part of the Patterson map within, say, 5–6 Å of the origin

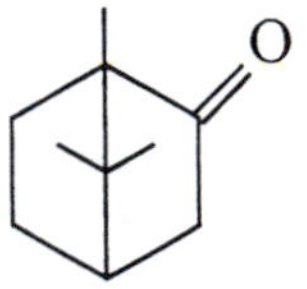

Fig. 8.2.   Molecular structure of camphor.

is considered, since few vectors in this region will be intermolecular ones. The calculated model for the known group is placed at the origin of the calculated Patterson function and systematically rotated about three axes in such a way as to give a fairly even grid in 'rotation space'. Various methods are used for this, one of which is analogous to the Eulerian angles used in diffractometer geometry, the angles being called A, B and C (A corresponds to ω, B to χ, and C to ϕ). When $B = 0$, A and C are rotations about the same axis, while when $B = \pi/2$ radians, A and C are rotations about mutually perpendicular axes. In practice, the search is not so large as might be expected.

Since Patterson maps are centrosymmetric, only half of the vectors need to be considered, and it is a property of the Eulerian system that the rotation A, B, C is identical in its effect to the rotation $\pi + A, -B, \pi + C$, so only half of the sphere of rotation need be inspected in any case. In an orthorhombic group, the test molecule will always be present in four different orientations, and any one of these will do, so as little as one-eighth of the sphere need be examined. The asymmetric unit for rotation is given in Table 8.2 for an unsymmetrical model rotated in maps of the three commonest symmetries.

Table 8.2 Asymmetric ranges for Patterson rotation functions

	Range of A	Range of B	Range of C
Triclinic, $P\bar{1}$	$0-2\pi$	$0-\pi$	$0-2\pi$
Monoclinic, $P2/m$	$0-2\pi$	$0-\pi/2$	$0-2\pi$
Orthorhombic, $Pmmm$	$0-2\pi$	$0-\pi/2$	$0-\pi$

At each position, the product of the vector density of the rotated model and the observed Patterson map is calculated at a range of grid points, and these products are summed. The larger the value of this sum, the more probable is the orientation of the search fragment corresponding to the current orientation of the vector map. The maximum value for the sum of these products indicates a high degree of agreement. For some of the better fits, a second search is done at a finer grid, and the best orientations are selected for the next stage of the process.

8.2.2 Translation search

With an orientation chosen, the structure is solved for a crystal in space group $P1$, since any point may be selected randomly as an origin. In any other case, the position of the known fragment relative to the symmetry elements, or alternatively the position of the origin of the unit cell relative to the fragment, must be found. This may be done by stepping the atoms of the orientated search fragment through the unit cell, calculating symmetry-related ones, and comparing the implied intermolecular vectors with those in the Patterson map in a way analogous to that used for the rotation function. If the process is successful, a set of trial positions in the unit cell for the atoms of the search fragment has been obtained, and the structure can usually be developed from there. Again, the search process is often not as large as might be expected. In space groups like $P2_1$, where the origin is not fixed along one axis, only a two-dimensional search is required. For primitive triclinic, monoclinic and orthorhombic space groups, the space which must be explored is shown in Table 8.3.

In practice, the direct approach is seldom used, as equivalent calculations may be done much more simply. In the 'correlation' procedure, data are expanded (if necessary) to a triclinic set and structure factors are calculated for the orientated fragment for each of a set of symmetry-related reflections. The data can then be rearranged into Fourier

Table 8.3   Ranges for Patterson translation functions

Symmetry element	Dimensions of search	Boundary of 'unit cell'		
$\bar{1}$, 222, etc.	3	$0-a/2$	$0-b/2$	$0-c/2$
2 or 2_1 (parallel to b)	2	$0-a/2$	–	$0-c/2$
m, a, c, n (perpendicular to b)	1	–	$0-b/2$	–

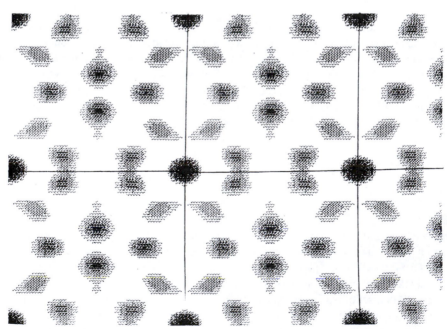

Fig. 8.3. The projection of an oxalate ion and its vector representation to the same scale as the Fourier and Patterson maps.

Fig. 8.4. Patterson map based on 19 data to a resolution of 2 Å for ammonium oxalate monohydrate.

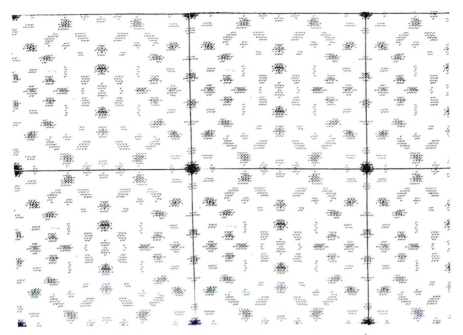

Fig. 8.5. Patterson map based on 120 data to a resolution of 1 Å.

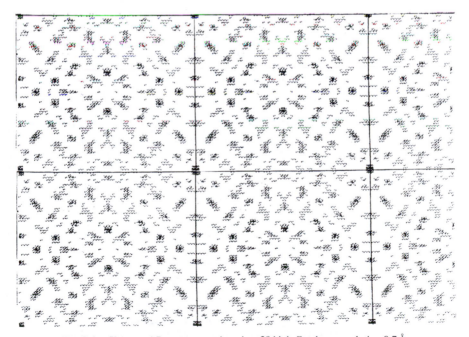

Fig. 8.6. Sharpened Patterson map based on 28 high E-values, resolution 0.7 Å.

coefficients for the sum $\sum F_o^2 F_c(t)^2$, where t represents the vector from the current origin to the actual one. It will have as many dimensions as required above for the search. This function will have maxima where many atomic vectors overlap and hence the highest point should correspond to the correct origin.

Representations of the Patterson function for the c-axis projection of ammonium oxalate are shown in Figs 8.4–8.6. These should be compared with those in Chapter 7, which were prepared from the same data. A vector pattern of the oxalate ion from the solved structure is also given on the same scale (Fig. 8.3). It was constructed using a piece of tracing paper by choosing a point as origin, placing the paper on the structure and moving it from atom to atom, marking the vectors to each other atom. The six (centrosymmetric) atoms result in six single-weight peaks, 12 double-weight peaks, and the sextuple-weight peak at the origin. It may be used to attempt a two-dimensional rotation function on the sharpest Patterson map (only one rotation is needed!).

Exercises

8.1 Calculate the Patterson vectors for a single heavy atom in the space group $P2_1/c$, the most common by far of all space groups. The general equivalent positions are:

$$x, y, z \quad -x, -y, -z \quad -x, \tfrac{1}{2}+y, \tfrac{1}{2}-z \quad x, \tfrac{1}{2}-y, \tfrac{1}{2}+z$$

8.2 A crystal in space group $P2_1/c$ has four Ru atoms per unit cell. The largest peaks in the Patterson map other than the origin peak are as follows.

u	v	w	Height
0.000	0.044	0.500	540
0.449	0.500	0.662	520
0.447	0.455	0.161	270

What other peaks will be related to these by the symmetry of the Patterson synthesis ($2/m$, C_{2h})? Calculate the position of any one Ru atom in the unit cell. The cell edges are all about 15 Å and β is close to 90°; can this compound be a chloro-bridged dimer?

8.3 A fragment for a Patterson search is shown in Fig. 8.7. The nickel atoms may be assumed to be octahedrally coordinated, and all the other atoms may be treated as S for this purpose, with an Ni–S bond length of 2.4 Å. Calculate a set of orthogonal Å coordinates for the atoms of the fragment.

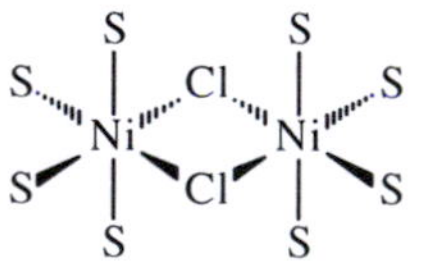

Fig. 8.7.   A Patterson search fragment for Exercise 8.3.

8.4 A structure is orthorhombic with eight heavy atoms in a C-centred unit cell. Conditions for reflections to be observed are: hkl, $h + k = 2n$ and $h0l$, $l = 2n$. Show that this is compatible with the space groups $Cmc2_1$, $C2cm$ (alternative setting of $Abm2$ with axes exchanged) and $Cmcm$. How could the Patterson function help to distinguish between the two non-centrosymmetric space groups? General equivalent positions are as follows:

$Cmc2_1$: (0,0,0) x, y, z $-x, -y, \frac{1}{2} + z$ $x, -y, \frac{1}{2} + z$ $-x, y, z$
and $(\frac{1}{2}, \frac{1}{2}, 0)+$

$C2cm$: (0,0,0) x, y, z $x, y, \frac{1}{2} - z$ $x, -y, \frac{1}{2} + z$ $x, -y, -z$
and $(\frac{1}{2}, \frac{1}{2}, 0)+$

$Cmcm$: (0,0,0) x, y, z $-x, -y, \frac{1}{2} + z$ $x, -y, \frac{1}{2} + z$ $-x, y, z$
and $(\frac{1}{2}, \frac{1}{2}, 0)+$ $-x, -y, -z$ $x, y, \frac{1}{2} - z$ $-x, y, \frac{1}{2} - z$ $x, -y, -z$

9

Direct methods of
crystal structure determination

Many methods of structure determination have been termed 'direct' (e.g. Patterson function, Fourier methods) in that, under favourable circumstances, it is possible to proceed in logical steps directly from the measured X-ray intensities to a complete solution of the crystal structure. However, the term 'direct' is usually reserved for those methods which attempt to derive the structure factor phases, electron density or atomic coordinates by mathematical means from a single set of X-ray intensities. Of these possibilities, the determination of phases is the most important for small-molecule crystallography.

9.1 Amplitudes and phases

The importance of phases in structure determination is obvious, but it is instructive to examine their importance relative to the amplitudes. To do this, we use the convolution theorem, which is set out in Appendix 1. It is not necessary to understand the mathematics in detail, but we are going to use the same relationship among the functions seen in equations (A1.35) and (A1.36).

Let us regard a structure factor as the product of an amplitude $|F(\mathbf{h})|$ and a phase factor $\exp(i\phi(\mathbf{h}))$, where $\mathbf{h}$ is a reciprocal space vector, the set of three indices being represented here by a single symbol. We will call the Fourier transform of $|F(\mathbf{h})|$ the 'amplitude synthesis' and the Fourier transform of the function $\exp(i\phi(\mathbf{h}))$ the 'phase synthesis'. The convolution theorem gives:

$$
\begin{array}{ccccc}
|F(\mathbf{h})| & \times & \exp(i\phi(\mathbf{h})) & = & F(\mathbf{h}) \\
\Updownarrow \text{F.T.} & & \Updownarrow \text{F.T.} & & \Updownarrow \text{F.T.} \\
\text{amplitude synthesis} & * & \text{phase synthesis} & = & \text{electron density}
\end{array} \tag{9.1}
$$

where $*$ is the convolution operator. The amplitude synthesis must look rather like the Patterson function with a large origin peak, and its convolution with the phase synthesis will put this large peak at the site of each peak in the phase synthesis. The phase synthesis must therefore contain peaks at atomic sites for the convolution to give the electron density. It is thus the phases rather than the amplitudes which give information about atomic positions in an electron density map. A good illustration of this was given by Ramachandran and Srinivasan [1] who calculated an electron density map using the phases from one structure (A) and the amplitudes from another (B). The map, in Fig. 9.1, shows the electron density peaks corresponding to the atomic positions in structure A

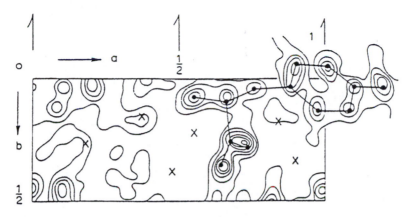

Fig. 9.1. Fourier synthesis calculated from $|F_\mathrm{B}| \exp(i\phi_\mathrm{A})$ where A and B are different structures. Atomic positions of structure A are marked by dots, those of structure B by crosses.

rather than B. Clearly, of the two problems nature could have given us, the phase problem is much more difficult than the amplitude problem.

9.2   The physical basis of direct methods

If the amplitude and phase of a structure factor were independent quantities, direct methods could not calculate phases from observed structure amplitudes. Fortunately, structure factor amplitudes and phases are not independent, but are linked through a knowledge of the electron density. Thus, if phases are known, amplitudes can be calculated to conform to our information on the electron density and, similarly, phases can be calculated from amplitudes. If nothing at all is known about the electron density, neither phases nor amplitudes can be calculated from the other.

However, something is always known about the electron density, otherwise we could not recognize the right answer when it is obtained. Characteristics and features of the correct electron density can often be expressed as mathematical constraints on the function $\rho(\mathbf{x})$ which is to be determined. Since $\rho(\mathbf{x})$ is related to the structure factors by a Fourier transformation, constraints on the electron density impose corresponding constraints on the structure factors. Because the structure amplitudes are known, most constraints restrict the values of structure factor phases and, in favourable cases, are sufficient to determine the phase values directly.

9.3   Constraints on the electron density

The correct electron density must always possess certain features like discrete atomic peaks (at sufficiently high resolution) and can never possess other features such as negative atoms. The electron density constraints that may be or have been used in structure determination are set out in Table 9.1. Constraints which operate over the whole cell are generally more powerful than those that affect only a small volume.

Table 9.1   Electron density constraints

Constraint	How used
1. Discrete atoms	Normalized structure factors
2. $\rho(\mathbf{x}) \geq 0$	Inequality relationships
3. Random distribution of atoms	Phase relationships and tangent formula
4. $\int \rho^3(\mathbf{x})\, dV = $ max.	Tangent formula
5. Equal atoms	Sayre's equation
6. $-\int \rho(\mathbf{x}) \ln(\rho(\mathbf{x})/q(\mathbf{x}))\, dV = $ max.	Maximum entropy methods
7. Equal molecules	Molecular replacement method
8. $\rho(\mathbf{x}) = $ constant	Density modification techniques

9.3.1   Discrete atoms

The first entry in the table, that of discrete atoms, is always available, since it is the very nature of matter. To make use of this information, we remove the effects of the atomic shape from the F_0 and convert them to Es, the normalized structure factors. The Es are therefore closely related to the Fourier coefficients of a point-atom structure. When they are used in the various phase-determining formulae, the effect is to strengthen the phase constraints so the electron density map should always contain atomic peaks. The convolution theorem shows the relationships among all these quantities:

$$
\begin{array}{ccccc}
E(\mathbf{h}) & \times & \text{atomic scattering factor} & = & F(\mathbf{h}) \\
\updownarrow \text{ F.T.} & & \updownarrow \text{ F.T.} & & \updownarrow \text{ F.T.} \\
\text{point atom structure} & * & \text{real atom} & = & \rho(\mathbf{x})
\end{array} \qquad (9.2)
$$

This relationship assumes all the peaks are the same shape, which is a good approximation at atomic resolution. The deconvolution of the map to remove the peak shape can therefore be expressed as

$$
|E(\mathbf{h})|^2 = \frac{|F_0(\mathbf{h})|^2}{\varepsilon_{\mathbf{h}} \sum_{i=1}^{N} f^2} \qquad (9.3)
$$

where $\varepsilon_{\mathbf{h}}$ is a factor which accounts for the effect of space group symmetry on the observed intensity. If the density does not consist of atomic peaks, this operation has no proper physical meaning.

9.3.2   Non-negative electron density

The second entry in Table 9.1 expresses the impossibility of negative electron density. This gives rise to inequality relationships among structure factors, particularly those of Karle and Hauptman [2]. Expressing the electron density as the sum of a Fourier series and imposing the constraint that $\rho(\mathbf{x}) \geq 0$ leads to the requirement that the Fourier coefficients, $E(\mathbf{h})$, must satisfy

$$
\begin{vmatrix}
E(0) & E(\mathbf{h}_1) & E(\mathbf{h}_2) & \ldots & E(\mathbf{h}_n) \\
E(-\mathbf{h}_1) & E(0) & E(-\mathbf{h}_1 + \mathbf{h}_2) & \ldots & E(-\mathbf{h}_1 + \mathbf{h}_n) \\
 & & \ldots & & \\
E(-\mathbf{h}_n) & E(-\mathbf{h}_n + \mathbf{h}_1) & E(-\mathbf{h}_n + \mathbf{h}_2) & \ldots & E(0)
\end{vmatrix} \geq 0. \tag{9.4}
$$

The left-hand side is a Karle–Hauptman determinant, which may be of any order, and the whole expression gives the set of Karle–Hauptman inequalities. Note that the elements in any single row or column define the complete determinant. These elements may be any set of structure factors as long as they are all different. Since the normalized structure factors $E(\mathbf{h})$ and $E(-\mathbf{h})$ are complex conjugates of each other, the determinant is seen to possess Hermitian symmetry, i.e. its transpose is equal to its complex conjugate.

An example of how the inequality relationship (9.4) may restrict the values of phases is given by the order three determinant

$$
\begin{vmatrix}
E(0) & E(\mathbf{h}) & E(\mathbf{k}) \\
E(-\mathbf{h}) & E(0) & E(-\mathbf{h} + \mathbf{k}) \\
E(-\mathbf{k}) & E(-\mathbf{k} + \mathbf{h}) & E(0)
\end{vmatrix} \geq 0. \tag{9.5}
$$

If the structure is centrosymmetric so that $E(-\mathbf{h}) = E(\mathbf{h})$, the expansion of the determinant gives

$$
E(0)[|E(0)|^2 - |E(\mathbf{h})|^2 - |E(\mathbf{k})|^2 - |E(\mathbf{h} - \mathbf{k})|^2]
$$
$$
+ 2E(\mathbf{h})E(-\mathbf{k})E(-\mathbf{h} + \mathbf{k}) \geq 0. \tag{9.6}
$$

The only term in this expression which is phase dependent is the last one on the left-hand side. Therefore, for sufficiently large Es, the inequality can be used to prove that the sign of $E(\mathbf{h})E(-\mathbf{k})E(-\mathbf{h} + \mathbf{k})$ must be positive.

It is instructive to see how this is expressed in terms of the electron density. Figure 9.2 shows the three sets of crystal planes corresponding to the reciprocal lattice vectors $\mathbf{h}$, $\mathbf{k}$ and $\mathbf{h} - \mathbf{k}$ drawn as full lines, the dashed lines being mid-way between. If the maxima of the cosine waves of the Fourier component $E(\mathbf{h})$ lie on the full lines, the minima will lie on the dashed lines and vice versa. If all three reflections $E(\mathbf{h})$, $E(\mathbf{k})$ and $E(\mathbf{h} - \mathbf{k})$ are strong and the electron density is to be positive, the atoms must lie in positions which are near maxima for all three components simultaneously. Examples of such positions are labelled A, B, C and D in Fig. 9.2. If most atoms in the structure are at positions of type A, $E(\mathbf{h})$, $E(\mathbf{k})$ and $E(\mathbf{h} - \mathbf{k})$ will all be strong and positive. If most atoms are at positions of type B, the three reflections will still be strong, but both $E(\mathbf{k})$ and $E(\mathbf{h} - \mathbf{k})$ will be negative. These results are set out in Table 9.2 together with signs obtained when most atoms are at sites of type C or D. In each case, it is seen that the product of the three signs is positive.

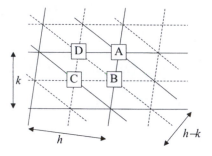

Fig. 9.2. The sign of $E(-\mathbf{h})E(\mathbf{h} - \mathbf{k})E(\mathbf{k})$.

Table 9.2 Signs of structure factors $E(\mathbf{h})$, $E(\mathbf{k})$ and $E(\mathbf{h}-\mathbf{k})$ when atoms are placed at the positions shown in Fig. 9.2

Position	s(h)	s(k)	s(h − k)	s(h)s(k)s(h − k)
A	+	+	+	+
B	+	−	−	+
C	−	−	+	+
D	−	+	−	+

A useful relationship of another type can be obtained from the order four determinant

$$\begin{vmatrix} E(0) & E(\mathbf{h}) & E(\mathbf{h}+\mathbf{k}) & E(\mathbf{h}+\mathbf{k}+\mathbf{l}) \\ E(-\mathbf{h}) & E(0) & E(\mathbf{k}) & E(\mathbf{k}+\mathbf{l}) \\ E(-\mathbf{h}-\mathbf{k}) & E(-\mathbf{k}) & E(0) & E(\mathbf{l}) \\ E(-\mathbf{h}-\mathbf{k}-\mathbf{l}) & E(-\mathbf{k}-\mathbf{l}) & E(-\mathbf{l}) & E(0) \end{vmatrix} \geq 0. \qquad (9.7)$$

Again, for a centrosymmetric structure and under the special conditions that $|E(\mathbf{h}+\mathbf{k})| = |E(\mathbf{k}+\mathbf{l})| = 0$, the expansion of the determinant gives the mathematical form

$$\text{terms independent of phase} - 2E(-\mathbf{h})E(-\mathbf{k})E(-\mathbf{l})E(\mathbf{h}+\mathbf{k}+\mathbf{l}) \geq 0. \qquad (9.8)$$

By cyclic permutation of the indices $\mathbf{h}$, $\mathbf{k}$ and $\mathbf{l}$, two similar determinants can be set up. Putting $|E(\mathbf{h}+\mathbf{k})| = |E(\mathbf{k}+\mathbf{l})| = |E(\mathbf{h}+\mathbf{l})| = 0$, and with large enough amplitudes for $|E(\mathbf{h})|$, $|E(\mathbf{k})|$, $|E(\mathbf{l})|$, $|E(\mathbf{h}+\mathbf{k}+\mathbf{l})|$, these can prove that the term $E(\mathbf{h})E(\mathbf{k})E(\mathbf{l})E(-\mathbf{h}-\mathbf{k}-\mathbf{l})$ must be negative.

9.3.3 Random atomic distribution

The constraint of non-negative electron density is only capable of restricting those phases which make $\rho(\mathbf{x})$ negative for some $\mathbf{x}$ for some value of the phase. This will be possible if the structure factor in question represents a significant fraction of the scattering energy. However, this is less likely to occur the larger the structure, and a point is soon reached where no phase information can be obtained for any structure factor. A more powerful constraint is therefore required which operates on the whole of the electron density no matter what its value.

This is achieved by combining the first two constraints in Table 9.1 to produce the third entry, where the structure is assumed to consist of a random distribution of atoms. The mathematical analysis gives a probability distribution for the phases rather than merely allowed and disallowed values. The probability distribution for a non-centrosymmetric structure equivalent to the inequality (9.6) is shown in Fig. 9.3 and is expressed mathematically as

$$P(\phi(\mathbf{h}, \mathbf{k})) = \frac{\exp[\kappa(\mathbf{h}, \mathbf{k}) \cos(\phi(\mathbf{h}, \mathbf{k}))]}{2I_0(\kappa(\mathbf{h}, \mathbf{k}))} \qquad (9.9)$$

where $\kappa(\mathbf{h}, \mathbf{k}) = 2N^{-1/2}|E(-\mathbf{h})E(\mathbf{h}-\mathbf{k})E(\mathbf{k})|$ and $\phi(\mathbf{h}, \mathbf{k}) = \phi(-\mathbf{h})+\phi(\mathbf{h}-\mathbf{k})+\phi(\mathbf{k})$. The number of atoms in the unit cell is N and $\phi(\mathbf{h})$ is the phase of $E(\mathbf{h})$, i.e. $E(\mathbf{h}) = |E(\mathbf{h})| \exp(i\phi(\mathbf{h}))$. It can be seen that the value of $\phi(\mathbf{h}, \mathbf{k})$ is more likely to be close to 0 than to π, giving rise to the phase relationship $\phi(-\mathbf{h})+\phi(\mathbf{h}-\mathbf{k})+\phi(\mathbf{k}) \approx 0$ (modulo 2π), i.e.

$$\phi(\mathbf{h}) \approx \phi(\mathbf{h}-\mathbf{k}) + \phi(\mathbf{k}) \qquad (9.10)$$

where the symbol '$\approx$' means 'probably equals'. The width of the distribution is controlled by the value of $\kappa(\mathbf{h}, \mathbf{k})$. Large values give a narrow distribution and hence a greater likelihood that $\phi(\mathbf{h}, \mathbf{k})$ is close to 0, i.e. tighter constraints on the phases.

9.3.4 Maximum value of $\int \rho^3(\mathbf{x}) \, dV$

The fourth entry in Table 9.1 is a powerful constraint. It clearly operates over the whole of the unit cell and not just over a restricted volume. It discriminates against negative

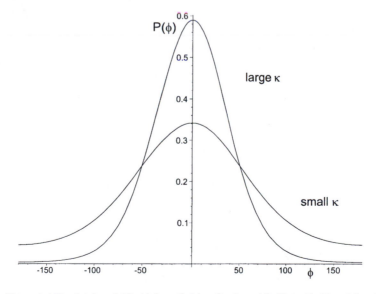

Fig. 9.3. The probability density of $\phi(\mathbf{h}, \mathbf{k})$ for $\kappa(\mathbf{h}, \mathbf{k}) = 2$, where $\phi(\mathbf{h}, \mathbf{k}) = \phi(-\mathbf{h}) + \phi(\mathbf{h}-\mathbf{k}) + \phi(\mathbf{k})$ and $\kappa(\mathbf{h}, \mathbf{k}) = 2N^{1/2}|E(-\mathbf{h})E(\mathbf{h}-\mathbf{k})E(\mathbf{k})|$.

density and encourages the formation of positive peaks, both expected features of the true electron density. This leads directly to probability relationships among phases and to the tangent formula, both of which have formed the basis of direct methods to the present day.

To obtain the tangent formula, the electron density in $\int \rho^3(\mathbf{x})\, dV$ is expressed as a Fourier summation, differentiated with respect to $\phi(\mathbf{h})$ and equated to zero to obtain the maximum. A rearrangement of the result gives

$$\tan(\phi(\mathbf{h})) \approx \frac{\sum_{\mathbf{k}} |E(\mathbf{k})E(\mathbf{h}-\mathbf{k})|\sin(\phi(\mathbf{k})+\phi(\mathbf{h}-\mathbf{k}))}{\sum_{\mathbf{k}} |E(\mathbf{k})E(\mathbf{h}-\mathbf{k})|\cos(\phi(\mathbf{k})+\phi(\mathbf{h}-\mathbf{k}))}. \tag{9.11}$$

This is expressed more concisely and with less ambiguity as

$$\phi(\mathbf{h}) \approx \text{phase of } \left\{ \sum_{k} E(\mathbf{k})E(\mathbf{h}-\mathbf{k}) \right\}. \tag{9.12}$$

Note that a single term in the tangent formula summation gives the phase relationship (9.10). Indeed, the tangent formula can also be derived from (9.9) by multiplying together the probability distributions with a common value of $\mathbf{h}$ and with different $\mathbf{k}$ and rearranging the result to give the most likely value of $\phi(\mathbf{h})$.

9.3.5   Equal atoms

The fifth entry in Table 9.1 has been included because of its extremely close relationship to the tangent formula in equation (9.11). In a large proportion of crystals, the atoms may be regarded as being equal. For example, in a crystal containing carbon, nitrogen, oxygen and hydrogen atoms only, the hydrogen can be ignored and the remaining atoms are approximately equal. This constraint was used by Sayre to develop an equation which gives exact relationships among the structure factors. If the electron density is squared, it will contain equal peaks in the same positions as the original density, but the peak shapes will have changed. This is expressed in terms of structure factors as

$$F(\mathbf{h}) = \frac{\Theta(\mathbf{h})}{V} \sum_{k} F(\mathbf{k})F(\mathbf{h}-\mathbf{k}) \tag{9.13}$$

where $\Theta(\mathbf{h})$ is related to the scattering factor of the squared atom.

Sayre's equation has had a profound influence on the development of direct methods. It is very closely related to the Karle–Hauptman inequalities and the tangent formula mentioned earlier. It is also used as a means of phase determination and refinement for macromolecules.

9.3.6   Maximum entropy

The sixth entry in Table 9.1 gives a measure of the entropy or information content of the electron density. It operates on the whole of the electron density and completely forbids negative regions. Maximizing the entropy of the electron density is a means of dealing with incomplete information, such as missing phases. A maximum entropy calculation produces a new map which has made no assumptions about the missing data and is therefore an unbiased estimate of the electron density, given all available information. This very promising technique has a number of important applications in crystallography, mainly in the area of structure refinement rather than structure determination. See Chapter 10 for further information on maximum entropy methods.

9.3.7 Equal molecules and $\rho(\mathbf{x}) = constant$

Entries 7 and 8 in Table 9.1 have both found use in macromolecular crystallography. When the same molecule occurs more than once in the asymmetric unit of a crystal, this immediately introduces the constraint that the electron density of the two molecules should be the same. The systematic application of this constraint constitutes the standard technique of molecular replacement in macromolecular crystallography.

There is little definite structure in the solvent regions of a macromolecular crystal, making the electron density almost constant outside the molecule. This information is exploited in the solvent flattening technique. This has been developed into a general density modification technique, which also includes Sayre's equation and other constraints and it is now in common use to improve the electron density maps of protein molecules.

9.4 Structure invariants

We have seen from the inequality relationship (9.6) that electron density constraints do not necessarily give the phases of individual structure factors. Instead, we obtain the value of a combination of phases. The same combination of phases occurs in the phase probability formula (9.9), the tangent formula (9.11) and in Sayre's equation (9.13).

The three structure factors are related such that the sum of their diffraction indices is zero and structure factor products which satisfy this criterion are known as structure invariants. Their special property is that the phase of the product is independent of origin position. The phase of a structure factor depends upon origin position, but its amplitude does not. Since only structure factor amplitudes feature in the phase-determining formulae, they can only define other quantities that are independent of origin position. Hence, the phase combination must be a structure invariant. In order to determine phases for individual structure factors, both the origin and enantiomorph must be defined first.

9.5 Structure determination

Direct methods of structure determination have become popular because they can be fully automated and are therefore easy to use. Now that the main formulae for phase determination have been presented, it remains to see how they are used in the determination of crystal structures. Computer programs which solve crystal structures in this way are readily available, and such a program will normally carry out the following operations:

(a) Calculation of normalized structure amplitudes, $|E(\mathbf{h})|$, from observed amplitudes $|F_o(\mathbf{h})|$. This is the rescaling of the F_o described in equation (9.3). It is normal also to use it to find the absolute scale of the Fs and produce intensity statistics as an aid to space group determination. Care must be taken in the estimation of Es for low-angle reflections.

(b) Set up phase relationships. Sets of three structure factors related as in equation (9.10) are identified and recorded for later use. Each such relationship is a single term in the tangent formula sum (9.11). Since this summation may be performed thousands of times, it is efficient to have all the terms already set up. In addition, four-phase structure invariants of the type seen in equation (9.8) are also set up for later use.

(c) Find the reflections to be used for phase determination. The phases of only the strongest $|E|$s can be determined with acceptable accuracy. In addition, each structure factor must be present in as large a number of phase relationships as possible. These two criteria are used to choose the subset of structure factors whose phases are to be determined.

(d) Assign starting phases. In order to perform the tangent formula summation (9.11), the phases of the structure factors in the summation must be known. Initially they may be assigned random values or phases calculated from an approximate electron density map.

(e) Phase determination and refinement. The starting phases are used in the tangent formula to determine new phase values. The process is then iterated until the phases have converged to stable values. With random starting phases, this is unlikely to yield correct phase values, so it is repeated many times as in a Monte Carlo procedure.

(f) Calculate figures of merit. Each set of phases obtained in (e) is used in the calculation of figures of merit. These are simple functions of the phases which can be calculated quickly and will give an indication of the quality of the phase set.

(g) Calculate and interpret the electron density map. The best phase sets as indicated by the figures of merit are used to calculate electron density maps. These are examined and interpreted in terms of the expected molecular structure by applying simple stereochemical criteria to the peaks found. Often, the best map according to the figures of merit will reveal most of the atomic positions.

9.5.1 Calculation of E values

Normalized structure amplitudes, $|E(\mathbf{h})|$s, are defined in equation (9.3), where $\varepsilon_\mathbf{h} \sum_{i=1}^{N} f^2$ is the expected intensity (also written as $\langle I \rangle$) of the $\mathbf{h}$ reflection. The best way of estimating $\langle I \rangle$ is as a spherical average of the actual intensities. In practice, the reflections are divided into ranges of $(\sin\theta)/\lambda$ and averages taken of intensity and $(\sin\theta)/\lambda$ in each range. Reflection multiplicities and also the effects of space group symmetry on intensities must be taken into account when the averages are calculated. Sampling errors can be decreased at low angles by using overlapping ranges of $(\sin\theta)/\lambda$.

Interpolation between the calculated values of $\langle I \rangle$ is aided if they can be plotted on a straight line, which is approximately true if a Wilson plot is used. Interpolation between the points on the plot can be done quite satisfactorily by fitting a curve locally to three or four points. This is repeated for different sets of points along the plot.

For best results, it is essential that the interpolated values of $\langle I \rangle$ follow the actual calculated points even if these depart very much from a straight line. Special care must be taken in calculating Es at low angles. If these are systematically overestimated this could easily result in failure to solve apparently simple structures. These Es are normally involved in more phase relationships than other reflections and therefore have a strong influence on phase determination. The number of strong Es chosen for phase determination is normally about $4 \times$ (number of independent atoms) $+ 100$. More than this may be needed for triclinic or monoclinic crystals.

9.5.2 Setting up phase relationships

Care should be taken to restrict the search for phase relationships to the unique ones only. However, in space groups other than triclinic, the same relationship may be set up more than once because of the symmetry operations. Such symmetry-related relationships should be summed so that the tangent formula automatically gives the correct symmetry phase restrictions. Normally about 15 times as many phase relationships as reflections should be found. If there are less than 10 times as many, more can be set up by including a few extra reflections. The number of three-phase relationships set up is roughly proportional to the cube of the number of Es used.

9.5.3 Finding reflections for phase determination

The phases of only the largest Es are usually determined and not all of these can be determined with acceptable reliability. It is therefore useful at this stage to eliminate about 10% of those reflections whose phases are most poorly defined by the tangent formula. An estimate of the reliability of each phase is obtained from $\alpha(\mathbf{h})$:

$$\alpha(\mathbf{h}) = 2N^{-\frac{1}{2}} |E(\mathbf{h})| \left| \sum_{\mathbf{k}} E(\mathbf{k}) E(\mathbf{h} - \mathbf{k}) \right|. \tag{9.14}$$

The larger the value of $\alpha(\mathbf{h})$, the more reliable is the phase estimate. The relationship between $\alpha(\mathbf{h})$ and the variance of the phase, $\sigma^2(\mathbf{h})$, is given by

$$\sigma^2(\mathbf{h}) = \frac{\pi^2}{3} + 4 \sum_{n=1}^{\infty} \frac{(-1)^n}{n^2} \frac{I_n(\alpha(\mathbf{h}))}{I_0(\alpha(\mathbf{h}))} \tag{9.15}$$

and the standard deviation, $\sigma(\mathbf{h})$, is shown in Fig. 9.4. From (9.14) it can be seen that $\alpha(\mathbf{h})$ can only be calculated when the phases are known. However, an estimate of $\alpha(\mathbf{h})$ can be obtained from the known distribution of three-phase structure invariants (equation (9.9)). A sufficiently good approximation to the estimated $\alpha(\mathbf{h})$ is given by

$$\alpha_e(\mathbf{h}) = \sum_{\mathbf{k}} K_{\mathbf{hk}} \frac{I_1(K_{\mathbf{hk}})}{I_0(K_{\mathbf{hk}})} \tag{9.16}$$

where $K_{\mathbf{hk}} = 2N^{-\frac{1}{2}} |E(\mathbf{h}) E(\mathbf{k}) E(\mathbf{h} - \mathbf{k})|$.

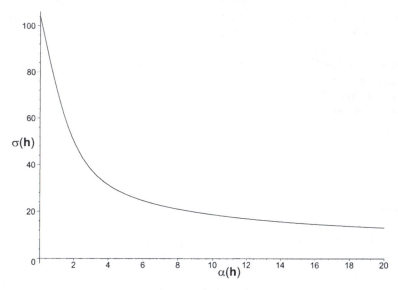

Fig. 9.4. The standard deviation of a calculated phase as a function of $\alpha(\mathbf{h})$.

The reflections with the smallest values of $\alpha_e(\mathbf{h})$ can now be eliminated in turn until the desired number remain.

9.5.4 Assignment of starting phases

All of the phases to be determined are assigned initial random values, which also serve to define the origin and enantiomorph of the subsequent electron density. It is not expected that starting phases assigned in this way will always lead to a correct set of phases after refinement, so the procedure is repeated a number of times as in a Monte Carlo technique. The number of such phase sets is normally between 30 and 200, but many more (or fewer) may be needed for some structures. Only one of these needs to be correct (and identified) for the structure to be solved. It may sometimes help if the starting phases are calculated from a random atomic distribution or perhaps one containing parts of the molecule available from a previous calculation, thus starting closer to the correct answer than a purely random guess.

9.5.5 Phase determination and refinement

The tangent formula and associated variance (9.15) are only correct under the assumption that the phases used in the calculation are correct. This is normally far from the truth. A crude attempt at correcting for this is to weight the terms in the summation so the tangent formula becomes

$$\phi(\mathbf{h}) = \text{phase of} \sum_{\mathbf{k}} w(\mathbf{k})w(\mathbf{h} - \mathbf{k})E(\mathbf{k})E(\mathbf{h} - \mathbf{k}) \tag{9.17}$$

where $w(\mathbf{h})$ is the weight associated with $\phi(\mathbf{h})$. The correct weight is inversely proportional to the variance and, to an adequate approximation, this is proportional to $\alpha(\mathbf{h})$
defined in (9.14).

A further improvement to the tangent formula is to include additional terms whose
most likely phase is π. These are the non-centrosymmetric equivalent of the relationship
(9.8) and are known as 'negative quartets'. They prevent all phases from refining to
zero in space groups such as $P1$ which contain no translational symmetry elements. The
modified formula is

$$\phi(\mathbf{h}) = \text{phase of } \{\alpha(\mathbf{h}) - g\eta(\mathbf{h})\} \tag{9.18}$$

$$\text{with } \eta(\mathbf{h}) = N^{-1}|E(\mathbf{h})|\sum_{\mathbf{kl}} E(-\mathbf{k})E(-\mathbf{l})E(\mathbf{h}+\mathbf{k}+\mathbf{l}).$$

$\alpha(\mathbf{h})$ is defined in (9.14) and g is an arbitrary scale factor to balance the effect of the two
terms $\alpha(\mathbf{h})$ and $\eta(\mathbf{h})$. The terms in the η summation are chosen such that the amplitudes
$|E(\mathbf{h}+\mathbf{k})|$, $|E(\mathbf{k}+\mathbf{l})|$ and $|E(\mathbf{h}+\mathbf{l})|$ are all extremely small or zero.

9.5.6   Figures of merit

The correct set of phases needs to be identified among the large number of incorrect
phase sets. This is done by figures of merit, which are functions of the phases that can
be rapidly calculated to give an indication of their quality. Among the most useful are
the following.

(a)
$$R_\alpha = \frac{\sum_{\mathbf{h}} |\alpha(\mathbf{h}) - \alpha_e(\mathbf{h})|}{\sum_{\mathbf{h}} \alpha_e(\mathbf{h})}. \tag{9.19}$$

This is a residual between the actual and the estimated αs. The correct phases should
make R_α small, but so do many incorrect phase sets. This is a better discriminator
against wrong phases which make R_α large.

(b)
$$\psi_0 = \frac{\sum_{\mathbf{h}} |\sum_{\mathbf{k}} E(\mathbf{k})E(\mathbf{h}-\mathbf{k})|}{\sum_{\mathbf{h}} (\sum_{\mathbf{k}} |E(\mathbf{k})E(\mathbf{h}-\mathbf{k})|^2)^{1/2}}. \tag{9.20}$$

The summation over $\mathbf{k}$ includes the strong Es for which phases have been determined
and the indices $\mathbf{h}$ are given by those reflections for which $|E(\mathbf{h})|$ is very small. The
numerator should therefore be small for the correct phases and will be much larger if
the phases are systematically wrong. The denominator normalizes ψ_0 to an expected
value of unity.

(c)
$$\text{NQUAL} = \frac{\sum_{\mathbf{h}} \alpha(\mathbf{h}) \cdot \eta(\mathbf{h})}{\sum_{\mathbf{h}} |\alpha(\mathbf{h})\| \eta(\mathbf{h})|}. \tag{9.21}$$

NQUAL measures the consistency between the two summations in (9.18) and should
have a low value for good phases. Correct phases are expected to give a value of -1.

To enable the computer to choose the best phase sets according to the figures of merit, a combined figure of merit is normally calculated. This is a sum of the scaled versions of the separate figures of merit and is usually the best indicator of good phases.

9.5.7 Interpretation of maps

Electron density maps are calculated using the best sets of phases as indicated by the figures of merit. The Fourier coefficients of these are normally Es rather than Fs because these are more readily available at this stage and they give sharper peaks. The slight disadvantage is that they also give a noisier background to the map. However, E-maps are usually preferred over F-maps.

Peaks in the maps should correspond to atomic positions but, because of systematic errors in the phases, there may be spurious peaks or no peaks where some atoms should be. It is normally sufficient to apply simple stereochemical criteria to identify chemically sensible molecular fragments. These may be displayed in a plot of peak positions on the least squares plane of the molecule, from which most of the molecule will be recognized.

9.5.8 Completion of the structure

If some atoms are missing from the map, the standard method of finding them is to use Fourier refinement (see Chapter 7). Phases calculated from the known atoms are used with weighted amplitudes to obtain the next map. Usually one or two iterations of this are sufficient to complete the structure. An alternative is to make use of all the diffraction data in Sayre's equation together with density modification to improve the map. This is normally used on macromolecular maps, but it is also very successful for small molecules. It will even convert an uninterpretable map into one in which most of the structure can be seen and the advantage is that it is all done automatically by the computer.

References

[1] G. N. Ramachandran and R. Srinivasan, *Nature*, 1961, **90**, 159.
[2] J. Karle and H. Hauptman, *Acta Cryst.*, 1950, **3**, 181.
[3] J. H. Robertson, *Acta Cryst.*, 1965, **18**, 410.

Exercises

9.1 Set up the order three Karle–Hauptman determinant for a centrosymmetric structure whose top row contains the reflections with indices **0**, **h** and **2h**. Hence obtain a constraint on the sign of $E(2\mathbf{h})$. What is the sign of $E(2\mathbf{h})$ if $E(\mathbf{0}) = 3$, $|E(\mathbf{h})| = |E(2\mathbf{h})| = 2$?

9.2 Verify equation (9.8). What sign information does it contain under the conditions $E(\mathbf{0}) = 3$, $|E(\mathbf{h})| = |E(2\mathbf{h})| = 2$, $|E(\mathbf{h} - \mathbf{k})| = 1$?

9.3 Expand the order four Karle–Hauptman determinant for a centrosymmetric structure whose top row contains the reflections with indices **0**, **h**, **h** + **k** and **h** + **k** + **l** and for which $E(\mathbf{h} + \mathbf{k}) = E(\mathbf{k} + \mathbf{l}) = 0$. Interpret your expression in terms of the sign information to be obtained and under which conditions it occurs.

9.4 Compare the Karle–Hauptman determinants with the following reflections in the top row: **0, h, h + k, h + k + l; 0, k, k + l, k + l + h; 0, l, l + h, l + h + k.** Summarize the sign information they contain when $E(\mathbf{h})$, $E(\mathbf{k})$, $E(\mathbf{l})$, $E(\mathbf{h + k + l})$ are all strong and $E(\mathbf{h + k}) = E(\mathbf{k + l}) = E(\mathbf{l + h}) = 0$.

9.5 Symbolic addition applied to a projection. Ammonium oxalate monohydrate [3] gives orthorhombic crystals, $P2_12_12$, with $a = 8.017$, $b = 10.309$, $c = 3.735\,\text{Å}$ (at 30 K). The short c-axis projection makes this an ideal structure for study in projection, as there can be little overlap of atoms. Data for the projection have been sharpened to point atoms at rest (i.e. converted to E-values) and are shown in Fig. 9.5. Note the mm symmetry and the fact that data are only present for $h00$ and $0k0$ for even orders, consistent with the screw axes. Find the especially strong data 5,7; −14, 5; 9,−12, which have indices summing to zero, as an example of a triple phase relationship (we omit the l index, since it is always zero for these reflections).

The problem is that phases must be assigned to the structure factors before they can be added up. Since this projection is centrosymmetric, phases must be 0 or π radians (0° or 180°), i.e. E must be given a sign + or −, but there are 2^{28} combinations of these values, and your chance of getting an interpretable map is small! Fortunately, the planes giving strong $|E|$ values are related by enough relationships to give us a unique, or almost unique,

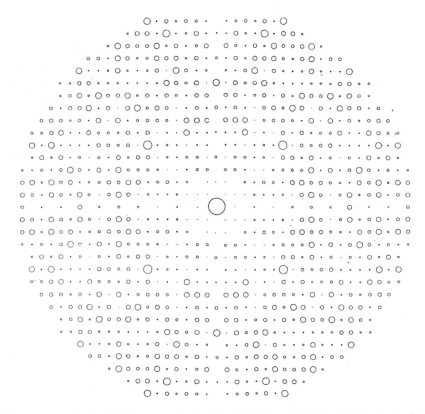

Fig. 9.5. c-Axis projection data for ammonium oxalate monohydrate. The index h runs vertically from the centre, and k runs horizontally.

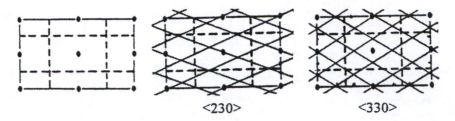

<230>   <330>

Fig. 9.6. Plane group symmetry for the ammonium oxalate monohydrate structure projection, together with two sets of lines (equivalent to planes in three dimensions).

solution. The main relationship used is that for large values of $|E|$, say $|E_1|$, $|E_2|$ and $|E_3|$ all > 1.5, if: $h_1 + h_2 + h_3 = k_1 + k_2 + k_3 (= l_1 + l_2 + l_3) = 0$, then: $\phi_1 + \phi_2 + \phi_3 \approx 0$. Additional help is given by the symmetry of the structure, illustrated in Fig. 9.6.

The plane group (two-dimensional space group) is *pgg*, with glide lines perpendicular to both axes, and there are four alternative positions for the origin: $0,0$; $0,\frac{1}{2}$; $\frac{1}{2},0$; and $\frac{1}{2},\frac{1}{2}$. This means that two phases may be arbitrarily fixed from any two of the parity groups g, u; u, g; or u, u (g and u mean even and odd, respectively, for the indices h and k), since, for example, shifting the origin by half a unit cell along a will shift the phase of all structure factors with h odd by π. Another result of the symmetry is that planes with indices h,k are related to $h,-k$ or $-h,k$ by the glide lines. The structure amplitudes must be the same for these, and the phases must be related, although they are not always the same. If h and k are both even or both odd, $\phi(h, k) = \phi(-h, k)$. If, however, one is odd and one even, $\phi(h, k) = \pi + \phi(-h, k)$. See the examples given for (23) and (33) in the diagrams. In other words, if we have a sign for a particular reflection h, k and we want the sign for either $-h, k$ or $h, -k$, then we must change the sign if $h + k$ is odd, but not if $h + k$ is even. Such sign changes are marked * in the list below.

					Determined signs
To get started, assign arbitrary signs to 5,7 and 14,5, and give 8,8 the					1 10
symbol A (unknown, to be determined). Data marked * have opposite					
signs to those which have both indices positive. Triples are arranged					1 14
from left to right and downwards in order of decreasing reliability.					
Note $A^2 = 1$ whatever the sign of A. For brevity, use B to stand for $-A$.					1 17

								Determined signs
5	7	−5	7	5	7	14	5	2 17
5	−7	14	5	10	0	−9	12*	
10	0	9	12	15	7	5	17	2 19
5	17	−5	7	14	−5*	5	7	3 10
5	−17	8	8	−8	8	6	−3*	
10	0	3	15	6	3	11	4	3 15
−5	7	9	−12*	5	17	−5	17	5 7
6	3	−3	15	6	−3*	6	−3*	
1	10	6	3	11	14	1	14	5 17

11	14	−1	14*	−1	10*	14	5	5 19
−10	0	10	0	8	−8	−7	−2	
1	14	9	14	7	2	7	3	6 3
5	7	−5	17	11	−4*	14	5	7 2
7	3	7	2	1	14	−9	14*	
12	10	2	19	12	10	5	19	7 3
5	19	11	−4*	−3	15	9	9	7 10
5	−19	1	10	12	−6	−8	8	
10	0	12	6	9	9	1	17	8 8
6	3	6	3	−5	7	9	−12*	8 13
6	3	7	3	13	6	1	17	
12	6	13	6	8	13	10	5	9 9
−3	15	−5	7	5	−7	−7	10*	9 12
10	−5*	7	10	−2	17*	10	0	
7	10	2	17	3	10	3	10	9 14
10	5	9	−9	5	19	−5	19	10 0
−9	9	−2	19*	2	−17*	13	−6*	
1	14	7	10	7	2	8	13	10 5
−2	17*	−1	−10					11 4
9	−14*	8	13					
7	3	7	3					11 14
								12 6
								12 10

The probable value for symbol A is:　　13 6

14 5

15 7

General bibliography

J. D. Dunitz, *X-ray Analysis and the Structure of Organic Molecules*. Cornell University Press, 1979.

C. Giacovazzo (ed.), *Fundamentals of Crystallography*, Oxford University Press, 1992.

C. Giacovazzo, *Direct Phasing in Crystallography*, Oxford University Press, 1998.

H. A. Hauptman, The phase problem of X-ray crystallography. In *Reports on Progress in Physics*, 1991, pp. 1427–1454.

M. F. C. Ladd and R. A. Palmer, *Theory and Practice of Direct Methods in Crystallography*, Plenum, 1980.

M. M. Woolfson, *Acta Cryst.*, 1987, **A43**, 593.

M. M. Woolfson, *An Introduction to Crystallography*, 2nd edn, Cambridge University Press, 1997.

10

An introduction to maximum entropy

When crystallographers talk about maximum entropy, they are usually referring to a technique for extracting as much information as possible from incomplete data. The missing data could be structure factor phases or perhaps some intensities where they overlap in a powder diffraction pattern. In this chapter we look at the ideas behind the technique.

10.1 Entropy

Entropy is a concept used in thermodynamics to describe the state of order of a system. A large body of mathematics has grown up around it, and since the same mathematics occurs elsewhere in science, the vocabulary of entropy has gone with it. Apart from thermodynamics itself, the fields to benefit most from these ideas are:

1. Information theory: entropy measures the amount of information in a message. The lower the entropy, the more information there is.
2. Probability theory: entropy measures the change in probability upon altering the conditions under which the probability is estimated. A low value of entropy corresponds to extremes of probability.
3. Image processing: entropy measures the amount of information in an image.

An increase in entropy means going from a less likely state to a more likely. Examples of increasing entropy may be that the temperatures of two bodies become more equal upon thermal contact, a message becomes slightly garbled upon transmission, or probabilities become less extreme because the information on which they are based has become outdated. In each case, you have to add something to the system to reverse the natural trend and thus decrease the entropy. Entropy naturally increases.

An illustration of this is the unlikely event portrayed in Fig. 10.1. There are many more ways of arranging lumps of wood to produce an untidy pile than there are to produce a useful shed. A random rearrangement of the wood is therefore more likely to produce the high entropy pile than the low entropy shed.

10.2 Maximum entropy

When entropy is maximized, it implies that all information has been removed: the message tells you nothing, everything has the same probability, and the image is completely flat. This is a fairly useless state to be in, but that is not the way in which maximum entropy is used as a numerical technique.

Fig. 10.1. An illustration of decreasing entropy.

Consider the application of maximum entropy to help form an image of a crystal structure, i.e. to produce as good an electron density map as the data will allow. Usually the data are both inaccurate (experimental error) and incomplete (low resolution, no phases, overlapped reflections), giving the possibility of an infinite number of maps which are consistent with the observed diffraction pattern.

How do you generate an acceptable map out of the infinite number of possibilities? Maximum entropy tries to do this by demanding that the map contains as little information as possible, i.e. its entropy is at a maximum, subject to the constraints imposed by the experimental data. This means that whatever information the map does contain is demanded by the data and is not there as a by-product of the numerical method or a hidden assumption. It also means that it can make no assumptions at all about the missing information and so produces as unbiased an estimate of the true map as possible.

10.3 Calculations with incomplete data

To illustrate how to deal with incomplete data, let us imagine we have the following information:

1. One third of all scientists make direct use of crystallographic data—let us call them crystallographers;
2. One quarter of all scientists are left-handed.

Table 10.1 The problem and some possible solutions

	(a) General statement		(b) Using the information		(c) Smallest maximum	
	lh	rh	lh	rh	lh	rh
Crystallographer	a	b	a	$\frac{1}{3}-a$	0	$\frac{4}{12}$
Non-crystallographer	c	d	$\frac{1}{4}-a$	$\frac{5}{12}+a$	$\frac{3}{12}$	$\frac{5}{12}$

	(d) Largest maximum		(e) Minimum variance		(f) Maximum entropy	
Crystallographer	$\frac{3}{12}$	$\frac{1}{12}$	$\frac{1}{24}$	$\frac{7}{24}$	$\frac{1}{12}$	$\frac{3}{12}$
Non-crystallographer	0	$\frac{8}{12}$	$\frac{5}{24}$	$\frac{11}{24}$	$\frac{2}{12}$	$\frac{6}{12}$

Now we pose the question: what proportion of all scientists are left-handed crystallographers?

The information given about left-handedness and crystallographic scientists is insufficient to answer the question precisely, but let us see how far we can go towards a sensible answer. The problem may be set out as in Table 10.1(a), where a is the proportion of left-handed crystallographers, d is the proportion of right-handed other scientists, and so on.

The information tells us that:

$$a+b=\tfrac{1}{3} \qquad a+c=\tfrac{1}{4} \quad \text{and} \quad a+b+c+d=1. \tag{10.1}$$

Using these three equations, we can eliminate b, c and d, and put everything in terms of the single variable a as in Table 10.1(b).

If we sensibly disallow negative scientists, any value of a between 0 and $\frac{1}{4}$ will be a possible answer to the question. For example, $a = 0$ gives one extreme solution, shown in Table 10.1(c), in which there are no left-handed crystallographers at all. The other extreme is given by $a = \frac{1}{4}$ (Table 10.1(d)), where all non-crystallographers are right-handed. As neither of these is very likely, we need a sensible criterion to apply to produce a plausible answer.

Let us see if it is sensible to seek a least-squares solution, i.e. find the value of a which minimizes the variance of the entries in Table 10.1(b). Such a criterion will certainly avoid the extreme solutions we have already looked at. The variance about the mean is given by:

$$V = \left(a - \tfrac{1}{4}\right)^2 + \left(\tfrac{1}{12} - a\right)^2 + a^2 + \left(\tfrac{1}{6} + a\right)^2 \tag{10.2}$$

and the minimum of V occurs when $dV/da = 0$, giving $a = \frac{1}{24}$. These results are shown in Table 10.1(e). It appears from this that $\frac{1}{8}$ of all crystallographers are left-handed ($\frac{1}{3} \times \frac{1}{8} = \frac{1}{24}$), but we see that $\frac{5}{16}$ of the non-crystallographers are also left-handed ($\frac{2}{3} \times \frac{5}{16} = \frac{5}{24}$). Why should there be this difference? The original information did not indicate this, and it seems we have made some hidden assumption which has caused it.

It actually arose because of the inappropriate technique used to obtain the result; there is no good reason why all the probabilities should be as close together as possible. What we really need is a method of obtaining an unbiased estimate of the number of left-handed crystallographers. Since crystallographers are not expected to be any different from other scientists in this respect, it would be reasonable to suppose that $\frac{1}{4}$ of them are left-handed like the rest of the scientific population. In the absence of any further information therefore, the most plausible result should be that shown in Table 10.1(f) with $a = \frac{1}{12}$.

Previously it was claimed that maximum entropy gave an unbiased estimate from incomplete data, i.e. it made no hidden assumptions about missing information. Will the maximum entropy solution therefore correspond to our most plausible solution in Table 10.1(f)? The formula for the entropy of the quantities in Table 10.1(b) will be derived later (see equation (10.9)), but that does not stop us from using it here. Applying the formula to our problem gives

$$S = -a \log(a) - \left(\tfrac{1}{3} - a\right) \log \left(\tfrac{1}{3} - a\right) + \left(\tfrac{1}{4} - a\right) \log \left(\tfrac{1}{4} - a\right)$$
$$- \left(\tfrac{5}{12} + a\right) \log \left(\tfrac{5}{12} + a\right) \tag{10.3}$$

and the maximum occurs when $dS/da = 0$, giving $a = \frac{1}{12}$. Maximum entropy therefore does give the most plausible solution, already seen in Table 10.1(f).

10.4   Forming images

Maximum entropy has clearly solved the problem in a most satisfying way, but can it actually produce electron density maps? Each array of numbers in Table 10.1 could just as easily represent an electron density map as probabilities of left-handedness among the scientific population. However, the left-handed crystallographer problem was chosen to illustrate the method because all the constraints are in the same space as the number array. With an electron density map, information about the amplitudes is in reciprocal space. This is therefore at the wrong end of a Fourier transform as far as the map is concerned, which introduces complications in the mathematics. We have spared you this so you can see more clearly how the method behaves.

10.5   Entropy and probability

Part of the definition of entropy is that the entropy of a complicated system is the sum of the entropies of its separate parts. The state of order of a system, which is measured by entropy, has a certain probability of occurring. That is, for each value of entropy, there is a corresponding value of probability. This may be written: $S = f(P)$, where S is entropy, P is probability and f is the function relating them.

Now, let us consider a system of two parts with entropies S_1 and S_2 and corresponding probabilities P_1 and P_2. If these states are independent of each other, the probability of the combined system is $P_1 \times P_2$ while its entropy is $S_1 + S_2$. That is, the entropy of the whole is

$$S = S_1 + S_2 = f(P_1) + f(P_2) = f(P_1 \times P_2). \qquad (10.4)$$

Compare this relationship with a fundamental property of the logarithm that

$$\log(a) + \log(b) = \log(a \times b). \qquad (10.5)$$

It is clear from this that we can write for the entropy $S = \log(P)$ and ignore any constant factors which may multiply the log function.

10.6   Electron density maps

If we could work out the probability of an electron density map occurring, this would immediately give a measure of its entropy. Let us imagine building a two-dimensional map on a tray using grains of sand. The tray is divided into small boxes corresponding to the grid points at which the map is normally calculated, and the density is represented by the number of sand grains piled up in each box. Throwing the sand onto the tray at random will produce a map, though usually not a very good one. However, the number of ways in which the grains of sand can be arranged to produce the map is a measure of how likely it is to occur. If there is only one way of arranging the sand to produce the map, it will occur only very rarely with a random throw.

We now need to work out how many ways the sand can be arranged to give a particular map. Figure 10.2 shows how a one-dimensional map can be made from individual grains. Assume the sand consists of N identical grains and that the map is built up by putting in place one grain at a time. The first grain has a choice of N places to go. The next grain has $N - 1$ choices, so the two together can be placed in the map in $N \times (N - 1)$ different ways. The third grain has $N - 2$ places to go, giving $N \times (N - 1) \times (N - 2)$ combinations of positions for the three grains. Thus, it can be seen that all N grains can be arranged in $N!$ ways altogether.

Since the grains are identical, it does not matter how they are arranged in each box in the tray; only the *number* of grains in the box affects the shape of the map. If there are n_1 grains in the first box, they can be arranged in $n_1!$ different ways within the box without affecting the map. For example, Fig. 10.3 shows the 6 ($= 3!$) possible arrangements of three grains. Therefore, the $N!$ combinations for the map must be reduced by this factor leaving $N!/n_1!$ combinations. Each box can be treated in this way, so the final number

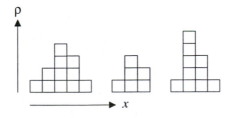

Fig. 10.2. A schematic representation of a one-dimensional map.

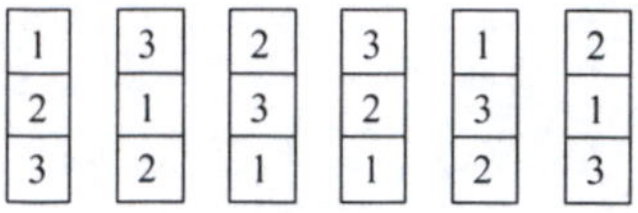

1	3	2	3	1	2
2	1	3	2	3	1
3	2	1	1	2	3

Fig. 10.3. Ways of arranging three objects in a box.

of different combinations of position for the grains of sand is:

$$\frac{N!}{n_1!n_2!\ldots n_m!} \tag{10.6}$$

where there are m grid points in the map.

This will be proportional to the probability of occurrence of the map. We can therefore obtain a measure of the entropy, S, by taking the log of this expression:

$$S = \log\left(\frac{N!}{n_1!n_2!\ldots n_m!}\right). \tag{10.7}$$

This is greatly simplified by making use of Stirling's approximation to the factorial of large numbers:

$$\log(N!) \approx N\log(N) - N. \tag{10.8}$$

Treating all the factorials in this way and remembering that $\sum n_i = N$ leads to the formula for the entropy of the map:

$$S = -\sum_{i=1}^{m} n_i \log(n_i). \tag{10.9}$$

If you measure electron density using electrons/Å^3 instead of counting grains of sand, the formula for entropy should be changed to

$$S = -\sum_{i=1}^{m} \rho_i \log\left(\frac{\rho_i}{q_i}\right) \tag{10.10}$$

where ρ_i is the density associated with the ith grid point and q_i is the expected density at that grid point. Initially, q_i will just be the mean density in the cell, but it can be updated as more information is obtained.

11

Least-squares fitting of parameters

In many scientific experiments, the experimental measurements are not the actual quantities required. Nearly always, the values of interest must be derived from those measured in the experiment. This is true in X-ray crystallography, where the atomic parameters need to be obtained from the X-ray intensities. It is this connection between parameters and experimental measurements which will be examined here.

11.1 Weighted mean

We will start with a simple situation in which only a single parameter is to be obtained, e.g. the length of a football pitch. A single measurement will not be sufficient, because it is easy to make mistakes, so we will measure it several times and take an average. Let us assume the measurements are 86.5, 87.0, 86.1, 85.9, 86.2, 86.0, 86.4 m, giving an average of 86.3 m. How reliable is this value? Is it really close to the true value or is it just a good guess? An indication of its reliability is obtained by calculating the variance about the mean:

$$\sigma^2 = \frac{1}{n-1} \sum_{i=1}^{n} (x_i - \bar{x})^2 \qquad (11.1)$$

where there are n measurements of value x_i and whose mean is $\bar{x}$. For the measurements of the football pitch, the variance σ^2 is $0.14\,\mathrm{m}^2$ and the standard deviation, σ, is about 0.4 m. For measurements which follow the Gaussian (normal) error distribution, there is a 68% chance of the true value being within one standard deviation of the derived value.

However, it may be possible to do better than this. We may know, for example, that the first two measurements were taken rather quickly and are less reliable than the others. It is sensible therefore to rely more on the good measurements by taking a weighted average:

$$\bar{x} = \frac{\sum_{i=1}^{n} w_i x_i}{\sum_{i=1}^{n} w_i} \qquad (11.2)$$

where the weights w_i are to be defined. The weights used should be those which give us the best value for the length of the pitch. However, this depends upon how we define 'best'. A very sensible definition, and the one most often used, is to define 'best' as that value which minimizes the variance. This leads directly to making the weights inversely proportional to the variance of individual measurements, i.e. $w_i \propto 1/\sigma_i^2$ and, for independent measurements, the variance will now be

$$\sigma^2 = \frac{n}{n-1} \frac{\sum_{i=1}^n w_i (x_i - \bar{x})^2}{\sum_{i=1}^n w_i}. \tag{11.3}$$

Minimizing the sum of the squares of the deviations from the mean gives the technique its title of 'least squares'.

To apply this to the length of the pitch, we must decide on the relative reliability of the measurements. Let us say we expect the error in the first two measurements to be about three times the error in the others. Since the variance is the square of the expected error, the weights w_1 and w_2 should be one-ninth of the other weights. Repeating the calculation using these weights gives 86.2 m for the length of the pitch with an estimated standard deviation of about 0.3 m. Note that the estimated length has changed and that the standard deviation is smaller.

11.2   Linear regression

A common example of the determination of two parameters is the fitting of a straight line through a set of experimental points. If the equation of the line is $y = mx + c$, then the parameters are the slope, m, and the intercept, c. The experimental measurements in this case are pairs of values (x_i, y_i), which may represent, for example, the extension of a spring, y_i, due to the force x_i. Lots of measurements can be taken and, for each measurement, we can write down an *observational equation* $mx_i + c = y_i$. This represents a system of linear simultaneous equations in the unknown quantities m and c in which there are many more equations than unknowns. There are no values of m and c which will satisfy the equations exactly, so we seek values which satisfy the equations as well as possible, i.e. give the 'best' straight line through the points on the graph.

The residual of an observational equation is defined as $\varepsilon_i = y_i - mx_i - c$. A common definition of 'best fit' is those values of m and c which minimize $\sum \varepsilon_i^2$ and this gives what statisticians call the line of linear regression. The recipe for performing the calculation is as follows. Let us write the observational equations in terms of matrices as

$$\begin{pmatrix} x_1 & 1 \\ x_2 & 1 \\ \vdots & \vdots \\ x_n & 1 \end{pmatrix} \begin{pmatrix} m \\ c \end{pmatrix} = \begin{pmatrix} y_1 \\ y_2 \\ \vdots \\ y_n \end{pmatrix} \tag{11.4}$$

or, more concisely, as

$$\mathbf{A}\mathbf{x} = \mathbf{b} \tag{11.5}$$

where, with a drastic change in notation, $\mathbf{A}$ is the left-hand-side matrix containing the x values, $\mathbf{x}$ is the vector of unknowns m and c, and $\mathbf{b}$ is the right-hand-side vector containing the y values. The matrix $\mathbf{A}$ is known as the *design matrix*. The least-squares solution of these equations is found by premultiplying both sides of (11.5) by the transpose of $\mathbf{A}$

$$(\mathbf{A}^T\mathbf{A})\mathbf{x} = \mathbf{A}^T\mathbf{b} \tag{11.6}$$

and solving the resulting equations for $\mathbf{x}$. These are known as the normal equations of least squares, which have the same number of equations as unknowns. This is, in fact, a general recipe. The observational equations (11.5) may consist of any number of equations and unknowns. Provided there are more equations than unknowns, the least-squares solution is obtained by solving the normal equations (11.6). The least-squares solution is defined as that which minimizes the sum of the squares of the residuals of the observational equations.

The observational equations can also be given weights. As in the calculation of the weighted mean, the weights, w_i, should be inversely proportional to the (expected error)^2 of each observational equation. The error in the equation is taken as the residual, so the correct weights are $\propto 1/(\text{expected residual})^2$. To describe mathematically how the weights enter into the calculation, we define a *weight matrix*, $\mathbf{W}$. It is a diagonal matrix with the weights as the diagonal elements and it premultiplies both sides of the observational equations (11.5), i.e.

$$\mathbf{WAx} = \mathbf{Wb}. \tag{11.7}$$

The normal equations of least squares are now

$$(\mathbf{A}^T\mathbf{WA})\mathbf{x} = (\mathbf{A}^T\mathbf{W})\mathbf{b} \tag{11.8}$$

which are solved for the unknown parameters $\mathbf{x}$. The weights ensure that the equations which are thought to be more accurate are satisfied more precisely. The quantity minimized is $\sum w_i \varepsilon_i^2$, where the w_i are the diagonal elements of $\mathbf{W}$.

11.3 Variances and covariances

Having obtained values for m and c, we now need to know how reliable they are. That is, how do we calculate their variances? Also, since there are two parameters, we need to know their covariance, i.e. how an error in one affects the error in the other. This is important for the calculation of any quantities derived from the parameters, such as bond lengths calculated from atomic positions. If a quantity x is calculated from two parameters a and b as

$$x = \alpha a + \beta b \tag{11.9}$$

then the variance of x is

$$\sigma_x^2 = \alpha^2 \sigma_a^2 + \beta^2 \sigma_b^2 + 2\alpha\beta\sigma_a\sigma_b\mu_{ab} \tag{11.10}$$

where $\sigma_a\sigma_b\mu_{ab}$ is the covariance of a and b, and μ_{ab} is the correlation coefficient. We can calculate both variances and covariances by defining a so-called *variance–covariance matrix*, $\mathbf{M}$. It contains the variances as diagonal elements and the covariances as off-diagonal elements. The matrix $\mathbf{M}$ is obtained as

$$\mathbf{M} = \frac{n}{n-p} \frac{\sum_{i=1}^{n} w_i \varepsilon_i^2}{\sum_{i=1}^{n} w_i} (\mathbf{A}^T \mathbf{W} \mathbf{A})^{-1} \tag{11.11}$$

where $(\mathbf{A}^T \mathbf{W} \mathbf{A})^{-1}$ is the inverse of the normal matrix of least squares and $\sum w \varepsilon^2 / \sum w$ is the weighted mean (residual)^2. This is a general recipe for the case where there are n observational equations with p parameters to be derived. The quantity $n - p$ is known as the number of degrees of freedom in the equations. In the case of linear regression, $p = 2$. Note that as p approaches the value of n, the variances increase. If $n = p$, i.e. there are as many equations as unknowns, no estimate of variance can be made using this recipe. To make the variances as small as possible, there should be many more equations than unknowns, i.e. $n \gg p$. This means that in crystallographic least-squares refinement, there should be many more observed reflections than parameters in the model.

11.4   Restraints

To illustrate some devices that are used in crystallographic least-squares refinement, let us see how inaccurate data can be treated in the following situation. Imagine that a totally unskilled surveyor measures the angles of a triangular field and gets the results $\alpha = 73°$, $\beta = 46°$, $\gamma = 55°$. He is so unskilled that he does not check to see if the angles add up to $180°$ until he gets back to the office, and by then it is too late to put things right. Can we do anything to help him? Obviously there is no substitute for accurate measurements, but we can always try to extract the maximum amount of information from the measurements we have.

There is the additional information already alluded to, that the sum of the angles must be $180°$. This information is not used in the measurement of the angles and so should help to correct the measurements in some way. If we included this as an additional equation and obtained a least-squares solution, would we get a better result? The system of equations will be:

$$\alpha = 73° \qquad \beta = 46° \qquad \gamma = 55°$$
$$\alpha + \beta + \gamma = 180° \tag{11.12}$$

and the least-squares solution is $\alpha = 74.5°$, $\beta = 47.5°$, $\gamma = 56.5°$. The effect of using the additional information is to change the sum of the angles from its original value of $174°$ to a more acceptable $178.5°$. This is called a *restraint* on the angles. Restraints are commonly used in least-squares refinement of crystal structures, such as when a group of atoms is known to be approximately planar or a particular interatomic distance is well known. Such information is included as additional observational equations. Note that all the equations (11.12) have the same residual of $1.5°$ when the least-squares solution is substituted in.

We may be able to help our hapless surveyor even more. Upon quizzing him, it emerges that the expected error in α is probably half that of the other measurements. This allows us to apply weights to the observational equations which are inversely proportional to the variance. The restraint also did not seem to be applied strongly enough. Perhaps we would like the sum of angles to be closer to $180°$ than it turned out to be, so let us also include the restraint with a larger weight. The weighted observational equations now look like this:

$$\begin{pmatrix} 4 & 0 & 0 & 0 \\ 0 & 1 & 0 & 0 \\ 0 & 0 & 1 & 0 \\ 0 & 0 & 0 & 4 \end{pmatrix} \begin{pmatrix} 1 & 0 & 0 \\ 0 & 1 & 0 \\ 0 & 0 & 1 \\ 1 & 1 & 1 \end{pmatrix} \begin{pmatrix} \alpha \\ \beta \\ \gamma \end{pmatrix} = \begin{pmatrix} 4 & 0 & 0 & 0 \\ 0 & 1 & 0 & 0 \\ 0 & 0 & 1 & 0 \\ 0 & 0 & 0 & 4 \end{pmatrix} \begin{pmatrix} 73 \\ 46 \\ 55 \\ 180 \end{pmatrix} \tag{11.13}$$

where the weight and design matrices have been written out separately. This time the least-squares solution gives $\alpha = 73.6°$, $\beta = 48.4°$, $\gamma = 57.4°$. It is seen that the sum of the angles is closer to $180°$ than before, i.e. $179.4°$, and α has moved away less from its measured value than the other angles, reflecting its greater accuracy.

However, this result deserves further comment. It was stated that the expected error in α is half that of the other angles, yet its shift in value is only one quarter of the shifts applied to β and γ. Why is this? The answer lies in an unjustified assumption which was made when setting up the weight matrix. The diagonal elements are all correctly calculated as inversely proportional to the variance of the corresponding equation, but the equations were all assumed to be independent of each other, making the weight matrix diagonal. The equations are certainly not independent, since the error in α in the top equation will also appear in an identical fashion in the bottom equation. Errors in β and γ appear similarly. This is correctly dealt with by taking into account the covariances of the equations, giving rise to off-diagonal elements in the weight matrix. In crystallographic least squares this is always ignored, so it will be ignored here also.

11.5   Constraints

What we should have done from the very beginning is to insist that the sum of the angles is exactly $180°$, which of course it is. Instead of using it as a restraint, which is only partially satisfied, we will now use it as a *constraint*, which must be satisfied exactly.

There are two standard ways of applying constraints and by far the most elegant is the following. If the equations we wish to solve are expressed as

$$\mathbf{Ax} = \mathbf{b} \tag{11.14}$$

and the variables $\mathbf{x}$ are subject to several linear constraints, then we can express the constraints as the equations

$$\mathbf{Gx} = \mathbf{f}. \tag{11.15}$$

In our example, the equations (11.14) are $\alpha = 73°$, $\beta = 46°$, $\gamma = 55°$ and there is only one constraint, given by the equation $\alpha + \beta + \gamma = 180°$. In the more general case, the solution of (11.14) subject to the constraints (11.15) is given by

$$\begin{pmatrix} \mathbf{A} & \mathbf{G}^{\mathrm{T}} \\ \mathbf{G} & \mathbf{0} \end{pmatrix} \begin{pmatrix} \mathbf{x} \\ \lambda \end{pmatrix} = \begin{pmatrix} \mathbf{b} \\ \mathbf{f} \end{pmatrix} \tag{11.16}$$

where the left-hand-side matrix consists of four smaller matrices as shown, and a vector of new variables λ has been introduced. There is one new variable for each constraint. The technical name for these additional variables is Lagrange multipliers. These equations are now solved for $\mathbf{x}$ and λ. Normally, the values of the λs are not required, so they can be ignored, and the xs are now the solution of (11.14) subject to the constraints (11.15).

Let us try this on the simple example of the angles in a triangle, but this time apply the sum of the angles as a constraint. Since this is no longer a least-squares calculation (there are the same number of equations as unknowns) the square root of the previous weights must be applied, giving the equations

$$2\alpha + \lambda = 146°$$
$$\beta + \lambda = 46°$$
$$\gamma + \lambda = 55°$$
$$\alpha + \beta + \gamma = 180°$$
(11.17)

The four equations are solved for α, β, γ and λ to give $\alpha = 74.2°$, $\beta = 48.4°$, $\gamma = 57.4°$ and $\lambda = -2.4°$. It can be seen that the constraint is satisfied exactly and that the shift in the value of α is half the shifts in β and γ, as you would expect.

If there are more equations than unknowns, as is normally the case, the equations (11.16) become

$$\begin{pmatrix} \mathbf{A}^T\mathbf{W}\mathbf{A} & \mathbf{G}^T \\ \mathbf{G} & \mathbf{0} \end{pmatrix} \begin{pmatrix} \mathbf{x} \\ \lambda \end{pmatrix} = \begin{pmatrix} \mathbf{A}^T\mathbf{W}\mathbf{b} \\ \mathbf{f} \end{pmatrix}$$
(11.18)

where a weight matrix has also been included. You may recognize in equation (11.18) the normal equations of least squares along with the constraint equations.

There are more equations in (11.17) than parameters we wish to evaluate. This is always the case when constraints are applied using Lagrange multipliers. In crystallographic least-squares refinement, the number of parameters is usually large (it can easily be several thousand) and any increase in the size of the matrix by the application of constraints is avoided if possible. Normally, crystallographers use a different method of applying constraints—one which will actually decrease the size of the matrix.

In this method, the constraint equations are used to give relationships among the unknowns so that some of them can be expressed in terms of others. In general, the constraint equations may be expressed as

$$\mathbf{x} = \mathbf{C}\mathbf{y} + \mathbf{d}$$
(11.19)

where $\mathbf{x}$ is the original vector of unknowns and $\mathbf{y}$ is a new set of unknowns, reduced in number by the number of independent constraints. Substituting this into (11.14) gives

$$\mathbf{A}\mathbf{C}\mathbf{y} = \mathbf{b} - \mathbf{A}\mathbf{d}$$
(11.20)

from which a least-squares solution for $\mathbf{y}$ is obtained. Since there are fewer unknowns represented by the $\mathbf{y}$ vector than those in $\mathbf{x}$, this is a smaller system of equations than we had before. The original variables $\mathbf{x}$ are then obtained from (11.19).

To illustrate this using the angles of a triangle, we can express γ in terms of α and β by the constraint $\gamma = 180 - \alpha - \beta$. We will call the reduced set of unknowns u and v such that

$$\begin{pmatrix} \alpha \\ \beta \\ \gamma \end{pmatrix} \begin{pmatrix} 1 & 0 \\ 0 & 1 \\ -1 & -1 \end{pmatrix} \begin{pmatrix} u \\ v \end{pmatrix} + \begin{pmatrix} 0 \\ 0 \\ 180 \end{pmatrix}, \tag{11.21}$$

and substituting for α, β, γ in the observational equations gives

$$\begin{pmatrix} 1 & 0 \\ 0 & 1 \\ -1 & -1 \end{pmatrix} \begin{pmatrix} u \\ v \end{pmatrix} = \begin{pmatrix} 73 \\ 46 \\ -125 \end{pmatrix}. \tag{11.22}$$

The normal equations of least squares are

$$\begin{pmatrix} 2 & 1 \\ 1 & 2 \end{pmatrix} \begin{pmatrix} u \\ v \end{pmatrix} = \begin{pmatrix} 198 \\ 171 \end{pmatrix}, \tag{11.23}$$

which give $u = 75°$, $v = 48°$, so that $\alpha = 75°$, $\beta = 48°$ and $\gamma = 57°$. No weights were used, so all angles are shifted by the same amount and their sum is exactly $180°$.

11.6   Non-linear least squares

At this point we discover that someone else in the surveyor's office has also measured the field, except he obtained the lengths of the three sides. These are $a = 21$ m, $b = 16$ m, $c = 19$ m. We can now do even better than before, because additional information is at hand. The sides are related to the angles using the sine rule. That is:

$$\frac{a}{\sin \alpha} = \frac{b}{\sin \beta} = \frac{c}{\sin \gamma} \tag{11.24}$$

which gives additional equations that can be added to our set. The difficulty is that the new equations are non-linear and, in general, there is no direct way of solving them to obtain values of the parameters. However, we need to be able to deal with these as well, since the equations for the refinement of crystal structures are also non-linear.

Let us use the new equations first of all as restraints. That is, they are simply added to the observational equations with appropriate weights. Our unweighted observational equations could now be:

$$\begin{array}{ccc} 2\alpha = 146° & \beta = 46° & \gamma = 55° \\ a = 21\,\text{m} & b = 16\,\text{m} & c = 19\,\text{m} \\ & a \sin \beta - b \sin \alpha = 0\,\text{m} & \\ & b \sin \gamma - c \sin \beta = 0\,\text{m} & \\ & \alpha + \beta + \gamma = 180° & \end{array} \tag{11.25}$$

i.e. nine equations in six unknowns. However, four of the equations give the value of an angle while the remaining five give a length. How can you compare length measurements with angle measurements? Does it matter whether you express the angles in terms of degrees or radians? This can all be taken care of in the weights assigned to the equations. A proper weighting scheme will make the expected variance of each weighted equation numerically the same and make sure that the expected errors in the parameters all have the same effect upon the equations. However, it is unnecessary to go into such details here. The same problem arises in the refinement of crystal structures where, for example, atomic displacement parameters are determined along with atomic positional parameters. Surprisingly, not every crystallographic least-squares program does this properly.

Using the current values of the sides and angles of the triangle will not satisfy the last three equations in (11.25). Our aim is to minimize the sum of the squares of the residuals of all the equations by adjusting the values of $a, b, c, \alpha, \beta, \gamma$, thus giving the least-squares solution. Now let us see how this least-squares solution may be obtained. Let the ith non-linear equation be $f_i(x_1, x_2, \ldots, x_n) = 0$ whose jth parameter is x_j. The derivative of f_i with respect to x_j is $\partial f_i/\partial x_j$. We can set up a matrix, $\mathbf{A}$, of such derivatives so the element a_{ij} is $\partial f_i/\partial x_j$. There will be as many rows in the matrix as observational equations and as many columns as parameters. We can also calculate the residual, ε_i, of each equation using the current parameter values. The shifts $\Delta\mathbf{x}$ to the parameters can then be calculated from the equations $\mathbf{A}\Delta\mathbf{x} = -\varepsilon$, where ε is the vector of residuals. When there are more equations than unknowns, least-squares values are obtained by solving

$$(\mathbf{A}^{\mathrm{T}}\mathbf{A})\Delta\mathbf{x} = -\mathbf{A}^{\mathrm{T}}\varepsilon. \tag{11.26}$$

The shifts to the parameters are then applied to give new values

$$\mathbf{x}_{\mathrm{new}} = \mathbf{x}_{\mathrm{old}} + \Delta\mathbf{x} \tag{11.27}$$

which should satisfy the observational equations better than the old values.

However, this recipe is strictly valid only for infinitesimally small shifts and is an approximation for parameter shifts of realistic size. This means the new parameter values are still only approximate and further shifts need to be calculated. A process of iteration is therefore set up, in which the latest parameter values are used to obtain new shifts and the operation is repeated until the calculated shifts are negligible.

Applying the recipe to the triangle example gives the matrix of derivatives

$$\begin{pmatrix}
1 & 0 & 0 & 0 & 0 & 0 \\
0 & 1 & 0 & 0 & 0 & 0 \\
0 & 0 & 1 & 0 & 0 & 0 \\
0 & 0 & 0 & 1 & 0 & 0 \\
0 & 0 & 0 & 0 & 1 & 0 \\
0 & 0 & 0 & 0 & 0 & 1 \\
-b\cos\alpha & a\cos\beta & 0 & \sin\beta & -\sin\alpha & 0 \\
0 & -c\cos\beta & b\cos\gamma & 0 & \sin\gamma & -\sin\beta \\
1 & 1 & 1 & 0 & 0 & 0
\end{pmatrix} \tag{11.28}$$

which is used to set up the equations for the parameter shifts.

Note that in order to apply this method of fitting the parameters to the experimental measurements, we need to begin with approximate values that are then adjusted to improve the fit with the experimental data. There is generally no way of determining these values directly from the non-linear equations.

11.7   Ill-conditioning

You are probably very familiar with the general rule that if anything can go wrong, it will. There are many traps for the unwary in least-squares refinement, but one that everyone must know about is ill-conditioning. Consider the innocent-looking pair of equations:

$$23.3x + 37.7y = 14.4 \qquad (11.29)$$

$$8.9x + 14.4y = 5.5$$

The exact solution is easily confirmed to be $x = -1$, $y = 1$. However, in the equations we normally deal with, the coefficients are subject to error—either experimental errors or errors in the model. Let us simulate a very small error in (11.29) by changing the right-hand side of the first equation from 14.4 to 14.39. Solving the equations this time yields the result $x = 13.4$, $y = -7.9$. The equations are, in fact, ill-conditioned and a very small change in the right-hand side has made the solution unrecognizably different. The computer will also introduce its own errors into the calculation, because it works to a limited precision. How can we trust the solution of a system of equations ever again? We clearly need to recognize ill-conditioning when we see it.

A common way of recognizing an ill-conditioned system of equations is to calculate the determinant of the left-hand-side matrix. In this case it is -0.01. An ill-conditioned matrix always has a determinant whose value is small compared with the general size of its elements. Another, related, symptom is that the inverse matrix has very large elements. In this case, the inverse is

$$\begin{pmatrix} -1440 & 3770 \\ 890 & -2330 \end{pmatrix}. \qquad (11.30)$$

Since the inverse matrix features in the formula (11.9) for calculating the variance–covariance matrix, an ill-conditioned normal matrix of least squares automatically leads to very large variances for the derived parameters.

In an extreme case, the determinant of the left-hand-side matrix may be zero. The matrix is then said to be singular and the equations no longer have a unique solution. They may have an infinite number of solutions or no solution at all. Physically, this means the equations do not contain the information required to evaluate the parameters. If the information is not there, there is no way of getting it from the equations. To make progress, it is necessary either to remove the parameters that are not defined or to add new equations to the system so that all the parameters *are* defined. There are a number of ways of producing a singular matrix in crystallographic least-squares refinement. Easy ways are to refine parameters that should be fixed by symmetry or to refine all atomic positional parameters in a polar space group; the symmetry does not define the origin

along the polar axis, so the atomic positions in this direction can only have relative
values.

11.8 Computing time

Most computing time in X-ray crystallography is spent on the least-squares refinement of
the crystal structure. An appreciation of where this time goes may help crystallographers
use their computing resources more efficiently.

The observational equations are mainly the structure factor equations, e.g.

$$\left| \sum_{j=1}^{N} f_j \exp[2\pi i(hx_j + ky_j + lz_j)] \right|^2 = |F_o(hkl)|^2 \tag{11.31}$$

which contain the atomic positional parameters for N atoms, and will usually also contain
the atomic displacement parameters in addition to occupancy factors, scale factors, etc.
There will be as many equations as observed structure factors; let this be n, with p
parameters describing the structure. If the calculation proceeds by setting up the normal
equations of least squares, this will be the most time-consuming part of the whole process.
The matrix multiplication alone (to form $\mathbf{A}^T\mathbf{A}$) will require about $np^2/2$ multiplication
operations. With p equal to a few hundred and n equal to a few thousand, $np^2/2$ will
typically be a few tens of millions ($\sim 10^7$). Efficient computer algorithms can reduce this
to a few times np operations, i.e. of the order of 10^6, but it is still large.

The amount of work required to solve the normal equations is about $p^3/3$ multiplications, while it takes about p^3 multiplications to invert the matrix. Note that it is
unnecessary to invert the matrix to solve the equations and so the matrix inverse should
only be calculated when it is needed in the estimation of variances. Again, computer
times may be reduced by using efficient algorithms, but matrix inversion will always
take a lot less time than that required to set up the normal equations.

Exercises

11.1   Show how equations (11.11) were derived and verify their least-squares solution.

11.2   Determine the slope and intercept of the line of linear regression through the points (1, 2),
        (3, 3), (5, 7), giving equal weight to each point.

11.3   Using data from Exercise 11.2, invert the normal matrix and, from this, calculate the
        correlation coefficient μ_{mc} between the slope m and intercept c.

11.4   In the triangle problem, let the expected errors in α, β, γ be in the ratio $1:2:1$.

        (a) Set up the weighted observational equiations for α, β, γ and include the restraint
            $\alpha + \beta + \gamma = 180°$ at half the weight of the equation $\alpha = 73°$.

        (b) Set up the normal equations of least squares from the observational and restraint
            equations.

        (c) Confirm that the solution of the normal equations is $\alpha = 73.6°$, $\beta = 48.4°$, $\gamma = 55.6°$.

11.5   In the triangle problem, let the observational equations be $\alpha = 73°$, $\beta = 46°$, $\gamma = 55°$,
        $a = 21\,\mathrm{m}, b = 16\,\mathrm{m}, c = 19\,\mathrm{m}$, and use the two restraint equations $a^2 = b^2 + c^2 + 2bc\cos\alpha$
        and $\alpha + \beta + \gamma = 180°$. Set up the matrix of derivatives needed to calculate shifts to the
        parameters.

12

Practical aspects of structure refinement

12.1 Introduction

From the outset we need to remember what is the purpose of least-squares refinement in crystal structure analysis. This is one of the most time-consuming parts of the whole process (certainly in computer CPU time, and often in human resources when it is not straightforward), and it is easy to get lost in the details and lose sight of the context. The overall aim is to find the description of the crystal structure, in terms of numerical parameters, which we believe best fits the evidence of the observed diffraction pattern. Once we have reliable estimates of reflection phases, which are rarely obtainable by direct experiment, Fourier transformation of the experimental reflection amplitudes and the estimated phases gives a three-dimensional electron density map, which is the primary structural result, in the sense that it is directly obtained from the data without the interpretation and approximations involved in a model based on atoms and bonds.

It is rare, however, for the results to be presented in terms of electron density. Usually the structure is described in terms of atom types, positions, and displacement parameters, together with estimates of precision of these numerical parameters ('standard uncertainties', s.u.s). The choice of parameters, and the exact way in which they are fitted to the experimental data, directly affect the result of the whole structure analysis and the way in which it is presented and interpreted. It is, therefore, a very important stage, despite the general view that it is not particularly interesting and the fact that it is often not well understood by those who use it.

The 'best fit' of structural parameters to the experimental data can be defined in various ways, and the traditional least-squares approach (itself appearing in a number of variants) is only one of these. In it, the calculated diffraction pattern corresponding to the numerically parameterized model structure (by Fourier transformation) is compared with the observed diffraction pattern. Since the experiment provides only amplitudes (or intensities) and not phases for the reflections, only these amplitudes can be compared. Altering the parameters of the model structure changes the calculated amplitudes or intensities; the observed amplitudes or intensities, of course, remain unchanged. The 'best fit' is defined as that which minimizes the function

$$\sum w(|F_o| - |F_c|)^2 \quad \text{or} \quad \sum w(F_o^2 - F_c^2)^2 \quad \text{or} \quad \sum w(I_o - I_c)^2 \qquad (12.1)$$

where each reflection has an assigned weight w which should represent our confidence in its importance in the sum, usually an estimate of the precision of the measured amplitude or intensity.

This particular method is widely used for crystal structure refinement because (a) it is based on a well-known, standard mathematical procedure; (b) it has been found to be reliable; and (c) it provides estimates of the precision of the refined parameters as well as the parameters themselves.

It does, however, have drawbacks. Compared to some other methods of refinement it is computationally expensive, and tricks to reduce the expense make it less reliable. Because the mathematical equations involved (Fourier transformation) are decidedly non-linear, an initial model structure is required, which must be reasonably close to the correct final answer, unlike the simple case of fitting a straight line to a set of points to obtain a gradient and intercept. It is subject to serious disturbance by individual poorly fitting data points ('outliers') and by inappropriately chosen weights. It is incapable, in itself, of suggesting additional parameters which should be added to the model. It can lead to a false 'local' minimum which does not represent the best possible fit of the available parameters to the data (though this is not common and can usually be recognized because of chemically unacceptable results and strange geometry). Some of these weaknesses are addressed by carrying out least-squares refinement and Fourier (or difference Fourier) map calculations together; electron density maps provide an assessment of the appropriateness of the model structure for the observed data, which does not directly depend on the particular numerical parameters used.

12.2   Data

The minimization function in least-squares refinement is

$$\sum w(Y_o - Y_c)^2 \tag{12.2}$$

where Y is usually either $|F|$ or F^2. There are sound statistical arguments for using $Y = I$, the raw intensities, and including as refinable parameters the various corrections normally applied in 'data reduction' to convert I to F^2, but this is rarely done, except for the common inclusion of one or more extinction parameters in the least-squares process. Lorentz-polarization (Lp) corrections are not dependent on the structural model, and refinement of absorption parameters together with anisotropic displacement parameters (ADPs) leads to problems of high correlation.

Fashions change in crystallography as in other spheres of life, and refinement on F^2 is currently in vogue, while refinement on F has fallen into relative disfavour. This is not just passing fashion, however, as there are some good, though not universally agreed, reasons for preferring F^2 refinement. In particular, it more easily allows all measured data to be used, suitably weighted. Inclusion of the very weakest reflections in F refinements is complicated by the difficulty of estimating $\sigma(F)$ from the experimental $\sigma(F^2)$ values (for use in weighting) when F^2 is zero or apparently negative, and by the question of what to do with negative measured intensities themselves, since a simple square root cannot be taken; omitting them or replacing them by zero or a small positive value introduces bias, which mainly affects ADPs. Weak reflections, especially when neighbouring reflections are strong, represent valuable information about the relative positions of atoms, and can be particularly important in cases of pseudo-symmetry. Certainly refinement against all F^2 values is superior to refinement against F values above a particular threshold. It also reduces the probability of settling into a local minimum, and makes the treatment of twinned structures and non-centrosymmetric structures simpler. There is, however, no point in using data at high angle if they are all weak, since these contain essentially no information.

Sometimes, especially for structures with substantial disorder, intensities fall off rapidly with increasing θ and there is a shortage of genuinely observed data. There is no doubt that low-temperature data collection almost always improves the precision and quality of data and reduces structure refinement problems. When there is a shortage of data, it is important not to attempt to refine too many parameters. The current guideline minimum data to parameter ratio in the Notes for Authors of *Acta Crystallographica Section C* is 8 for a non-centrosymmetric structure, 10 for centrosymmetric; with data collected to a resolution (d spacing) of 0.8 Å or better ($\theta_{max} > 25°$ for Mo Kα, 67° for Cu Kα), it should generally be easy to achieve a ratio between 10 and 20. Perhaps a better guide is the ratio of 'observed' data (those with $F^2 > 2\sigma(F^2)$, say) to refined parameters, which should preferably exceed 6. Note that symmetry-equivalent data should always be merged and not treated as independent for such calculations, but that Friedel opposites are distinct data for non-centrosymmetric structures where there is significant anomalous scattering.

The correct weighting of reflections is important for a successful refinement. In principle, the weight of each reflection should be inversely proportional to the estimated variance of its amplitude or intensity (whichever quantity is being used for refinement). Thus, in a refinement on F^2, $w = 1/\sigma^2(F^2)$ is appropriate, provided the $\sigma(F^2)$ values are correctly estimated. Unfortunately, $\sigma(F^2)$ values derived purely from Poisson counting statistics are an underestimate, because they ignore the effects of (known or unknown) systematic errors. It is common practice to inflate the variances, and hence reduce the weights, by adding term(s) depending on F^2 itself, and, possibly, other factors. This may be done as part of the data reduction process (expressions such as 'machine instability factor', 'ignorance factor' or even 'p-factor' are often used to describe this adjustment), or may be included in the refinement, with the extra terms as either fixed or variable parameters. Unit weights, treating all reflections equally, are never acceptable for final refinements with diffractometer data; they are particularly bad for F^2 refinements. Refinement on F^2 does depend heavily for its success on reasonable weights, and is much more sensitive to a poor weighting scheme and to outliers in the data. It is believed, however, to be less likely to lead to a false minimum. In the final analysis, both refinements, suitably weighted, should converge to essentially the same result.

The progress of refinement is directly monitored by the value of the minimization function, which should, obviously, reduce and converge to a minimum. Various 'residual factors' or R-indices are more commonly quoted, which provide a normalized version of the minimization function. The index most closely related to the minimization itself is

$$wR = \left[\frac{\sum w(Y_o - Y_c)^2}{\sum wY_o^2} \right]^{1/2} \tag{12.3}$$

which incorporates the reflection weights. Historically, an unweighted index has been most widely used:

$$R = \frac{\sum \|F_o| - |F_c\|}{\sum |F_o|} \tag{12.4}$$

based on $|F|$ values of 'observed' reflections. The advantages of calculating and quoting this index, together with wR, are that it provides a comparison with older work, it is generally considerably smaller than wR for F^2 refinements (a purely cosmetic advantage!), and it is relatively insensitive to changes in the weighting scheme, which can dramatically alter wR.

The goodness of fit is defined as

$$S = \left[\frac{\sum w(Y_o - Y_c)^2}{(N - P)} \right]^{1/2} \tag{12.5}$$

for N data and P refined parameters. In theory, if the weights are correct, S should be close to unity. In practice, this can always be fudged by rescaling the weights, which proves nothing, and a single value of this kind is, in any case, a poor way of assessing the quality of fit between the observed and calculated data. More important is that S (or some other related indicator) calculated for different groups of reflections (for example, ranges of reflection indices, Bragg angle, or F^2 values) should be fairly constant. Any systematic trends indicate either a fault in the model structure, a problem with the data, inappropriate weights, or a combination of these. Such an 'analysis of variance' can be used as a basis for making small adjustments to the weighting scheme, particularly the addition of F^2-dependent terms to the experimental $\sigma^2(F_o^2)$ values, but any unusual weighting functions which result from this should be regarded with grave suspicion, as they are almost certainly masking problems with the structure and/or data. The weighting scheme should not be adjusted on this basis until all desired parameters are included in the refinement, otherwise down-weighting poor agreements will suppress important information about shortcomings in the structural model. A list of the worst agreements between observed and calculated reflections will help to identify individual poor fits or 'outliers' which may be removed from the data set on the grounds of experimental error (such as a hardware fault or a reflection obscured by the beam stop, etc.), but this should not apply to more than a very few reflections.

12.3 Parameters

Parameters are the (mainly numerical) descriptors of the model structure. Those which are usually recognized and regarded as available for refinement in any structure are:

(a) *atomic coordinates*: three for every atom in a general position, but one or more will be constrained for an atom in a special position (see Chapter 3);
(b) *atomic displacement parameters*: one U (or B) value per atom for an isotropic model (equal displacement in all directions), six U values per atom for an anisotropic model (the commonly used ellipsoid representation with three principal axes in arbitrary

orientations; there will be fewer than six for some kinds of special positions, as shown in Chapter 3);

(c) **an overall scale factor** to bring the observed (arbitrary scale) and calculated ($F(000)$ is the number of electrons in the unit cell) amplitudes or intensities to the same scale.

There may be further scale factors, for example, if data were measured from more than one crystal. One or more parameters may be refined to deal with extinction effects. Additional parameters may be required for twinned structures and one parameter is refined by some programs for absolute structure determination of non-centrosymmetric structures. Thus, for most routine structures, the majority of refined parameters are ADPs.

Another atomic parameter which can be refined is the site occupancy. This is constrained to unity in most cases (i.e. not allowed to refine), but is available for manual adjustment or automatic refinement in cases of structural disorder, which is usually modelled by partial atoms. The number of refined parameters can rise dramatically for structures with substantial disorder.

It should be noted that atom types are also essentially structural parameters. They may not immediately appear to be numerical, but they enter the calculations as atomic scattering factor functions. These are not usually refined, but are assigned by the crystallographer on the basis of knowledge of the expected structure and electron densities found in calculated maps. A misassigned atom type will cause problems in the refinement, with refinable parameters adjusting in an attempt to compensate for the errors. Correlations among atom types, occupancy factors and ADPs are high.

There is also a very real sense in which the space group is a parameter of the model structure, although this is one which affects the treatment of the data (in the choice of portions of reciprocal space to be measured and in the merging of what are believed to be symmetry-equivalent reflections) as well as the refinement. In cases of unambiguous space group indications from the systematic absences of the data, the space group is usually fixed from the outset, but in other cases there may be more than one space group consistent with the data, and the different possibilities need to be explored during structure solution and refinement. Again, an incorrectly assigned space group can cause serious problems in refinement.

One of the decisions to be made during refinement is whether to treat hydrogen atoms like other atoms, with free refinement of their parameters, or to apply constraints and/or restraints to them. The decision depends on various factors, particularly the purpose of the experiment and the quality of the data, but the low electron density of hydrogen atoms always makes their refinement much less precise than that of other atoms, and the non-spherical distribution of the electron density around the nucleus introduces systematic displacements of the hydrogen atom positions derived from X-ray diffraction compared with those from neutron diffraction. In many cases, free refinement of hydrogen atoms gives no significant improvement in the fit, and produces geometry of low precision for these atoms, at the cost of a substantial increase in the number of refined parameters. It is not usually recommended unless the hydrogen atoms are of particular interest or are difficult to place geometrically, such as in cases of hydrogen bonding, aqua ligands, hydrido ligands, or bridging hydrogen atoms in boranes.

12.4   Constraints

Constraints and restraints (called 'hard constraints' and 'soft constraints' in some older literature) are devices which may be necessary in cases where parameters should be fixed to certain values, for example by symmetry, and which offer ways of assisting difficult refinements, for example those of disordered structures. The distinction between constraints and restraints is important.

A constraint is an exact mathematical relationship which affects one or more parameters in such a way that not all the parameters can be freely and independently refined. Its application reduces the number of refined parameters, either by fixing some or by tying some together, and it must be obeyed, whatever the cost to the rest of the refinement. It is absolutely rigid.

A restraint is an approximate target value for a particular parameter or a function of some parameters. It provides an extra piece of information about the structure, which is fed into the refinement alongside, and perhaps in competition with, the diffraction data. It thus increases the number of observations rather than reducing the number of parameters. Its balance against the demands of the diffraction data is flexible. Restraints are considered further in the next section.

The most commonly used constraints are on atom site occupancy factors. When there is no disorder in a structure, each atomic site is fully occupied, and the site occupancy factors are fixed. For a simple case of disorder in which one atom can occupy one of two alternative positions (such as a chloride anion disordered over two sites or a hydroxy group with its hydrogen atom disordered over two orientations), the two site occupancy factors can be refined, but they are not independent, since their sum is constrained to be unity; instead of two parameters, there is effectively only one, so that one site has occupancy s and the other $1 - s$. More complex cases of disorder involve constraints on groups of atoms, not just individual atoms.

Constraints are required for the coordinates of atoms on special positions. Examples are

(a) an inversion centre, for which all three coordinates are fixed;
(b) a mirror plane perpendicular to b, for which y is fixed;
(c) a twofold axis parallel to b, for which x and z are fixed;
(d) a body-diagonal threefold axis in a cubic space group, for which $x = y = z$.

ADPs also require constraints for some special positions. Examples are

(a) $U^{12} = U^{23} = 0$ for an atom on a mirror plane perpendicular to b;
(b) $U^{11} = U^{22}$ and $U^{12} = U^{13} = U^{23} = 0$ for an atom on a fourfold rotation axis in a tetragonal space group.

Note that an inversion centre does not impose any U^{ij} constraints.

In space groups with a floating origin (e.g. $P2_1$ with no defined origin for the b axis, and $P1$ for which the origin is arbitrary in all three axes), it is not possible to refine all the atomic coordinates freely, because this leads to a singular matrix. One commonly used remedy is to fix one (often the heaviest) atom to define the origin. This is simple, but it has the disadvantage that this particular atom then appears to be more precisely located than the others, and the s.u.s of the bond lengths and angles involving it are underestimated unless full account is taken of parameter covariances. A better method is to constrain the sum of coordinates of all the atoms to remain constant; this removes one parameter for each floating axis. Alternatively, the sum of coordinates can be restrained to a constant value with a high weight, which has more or less the same effect as a constraint, although the mathematics are quite different.

Constraints are often used in the refinement of hydrogen atoms. A 'riding model' is particularly effective, in which hydrogen atoms are positioned according to expected geometry and then the X–H vectors are constrained to remain constant in length and direction; the geometry idealization can be repeated after each refinement cycle. This procedure includes small but often significant X-ray scattering effects of hydrogen atoms in the calculations, but adds no extra positional parameters. Variants on this include allowing some restricted freedom, such as torsional freedom about the X–C bond for a methyl group, which adds just one parameter for three extra atoms.

The isotropic displacement parameters of hydrogen atoms can be fixed at predetermined values, tied together in groups (e.g. one U value for all methyl hydrogen atoms and another for all other hydrogen atoms, which probably vibrate less), or made to ride on the ADPs of the parent atoms, e.g. with $U_{iso}(H) = 1.2U_{eq}(C)$.

Another way of reducing the number of refined parameters is to put several atoms together in a rigid group, usually with idealized geometry. This has particular applications for badly disordered structures, in which the individual components overlap so that atoms are not resolved in the electron density, but the expected geometry is well known, as for some high-symmetry small counter-ions. In such cases, free refinement is often unsuccessful. A rigid group of N atoms has only six parameters, three translational and three rotational, instead of the $3N$ required for free refinement of all the atoms. Restraints offer an alternative approach for this situation.

Because constraints are enforced, even if they conflict seriously with the evidence of the diffraction data (they effectively have infinite weight), it is essential to choose appropriate ones. Unsuitable constraints will force other parameters to compensate and hence introduce inaccuracies into the resulting structure. Examples of unsuitable constraints are (a) an imposed C–H bond length of 1.08 Å, which is an expected internuclear distance, while the apparent C–H distance from X-ray diffraction is generally around 0.95 Å; and (b) a perfect hexagon as a rigid group for a phenyl substituent attached to an electronegative element, for which the angles at the *ipso*-C atom are usually significantly displaced from 120°.

12.5 Restraints

A restraint is treated as an experimental observation, just like the reflection amplitudes or intensities. The minimization function becomes

$$\sum w(Y_o - Y_c)^2 + \sum w(r_t - r_c)^2 \tag{12.6}$$

in which r_c is a quantity which can be calculated from the model structure parameters (e.g. a distance) and which is to be restrained to a particular target value r_t. Each restraint needs an associated weight w which indicates how strongly it is to be applied. If the relative weights of the diffraction data and the restraints are correct, the restraint weight is the inverse of the desired variance, $w = 1/\sigma^2$, which is easy to understand. Thus a restraint can be given a high weight (low σ) if we want it to be closely obeyed, and a low weight (high σ) if we want it to be just a weak preference. Restraints are, therefore, much more flexible than constraints for geometrical relationships among groups of atoms. The whole idea of restraints is to provide extra 'chemical structural' information to supplement the diffraction data when these are inadequate to define some of the model structure parameters properly. They should be used in cooperation with the diffraction data, not in direct competition with them. Such an adversarial approach is unlikely to produce an acceptable refined structure: conflicts between data and restraints inevitably mean increases in R-indices and final geometry s.u.s, or ineffective restraints.

In chemical crystallography, restraints are most often applied to geometry (combinations of atomic positional parameters) and to ADPs.

A simple example of a geometrical application is to restrain a particular bond length to a desired value. This may be necessary because disorder makes the position of one or both of the atoms rather imprecise. Such a distance restraint is not confined to bonds, but can be applied to any pair of atoms. As an extension, several distances may be restrained to be approximately equal, without actually specifying a numerical target value. Similarly, angles can be restrained, involving three atoms together. Flexible and powerful refinement programs allow the user to specify that two or more groups of atoms should have 'similar geometry' and this desire is translated into appropriate distance and angle restraints.

Other geometrical restraints include the wish to have a group of atoms lie approximately in a plane, and to have a particular approximate degree of pyramidality for an atom with three bonds ('chiral volume').

Two particularly useful restraints apply to ADPs. One restrains the components of ADPs along a bond between two atoms to be equal, and represents a rigid bond model. The other restrains ADPs to be similar for two atoms which are close together; this can help to prevent indeterminate parameters for disorder components which partially overlap.

Restraints will often not be needed for well-behaved refinements. They can be extremely helpful, however, in cases of disorder and pseudo-symmetry, and there should be no hesitation in exploiting them in such cases. The effectiveness of restraints can always be checked by comparing the desired numerical values with those which result from the refinement. Any marked deviations show that these restraints are being overruled by the diffraction data. In such cases, it may be appropriate to increase their weights, but their validity should be examined carefully.

12.6   Refinement procedures

For the refinement of P parameters from N data, each cycle involves building up a symmetric $P \times P$ matrix and a single array of length P; the individual data points and their mathematical derivatives with respect to each of the parameters contribute to the elements of the matrix and the array. After construction, the matrix is inverted and used to calculate parameter shifts and s.u.s. Such a process is called 'full-matrix least-squares refinement'.

Most of the off-diagonal terms of the least-squares matrix are usually considerably smaller than the terms on the leading diagonal. Ignoring them all (treating them as zero) gives a diagonal approximation to the method. A compromise involves calculating and using only the off-diagonal terms which are expected to be of significant size and ignoring the others. This is called the 'block-diagonal' approximation. It has the advantage of requiring less memory storage and less computing time than a full-matrix calculation. However, it loses information about correlations between parameters, and this usually means more cycles are required to reach convergence, and s.u.s are less reliable.

For really big structures (biological macromolecules), matrix methods become prohibitive, and alternative strategies are used, particularly the 'conjugate gradient' procedure. It is much faster per cycle, but more cycles are needed and no s.u.s are available at all. It is unlikely to be used for chemical structures, although it can be very useful as a fast method in the early stages of refinement of particularly large structures.

In the early stages of refinement, if the initial model structure is not very good, there may be no progress towards the correct answer. Sometimes the refinement 'blows up', with huge parameter shifts and massive R-indices. In such cases, imposition of constraints and restraints may help to increase the 'radius of convergence'; they can be relaxed or removed later. Another approach is to use 'shift-limiting restraints'. These restrict the amount by which parameters are allowed to change in one cycle, to avoid a catastrophic move away from the true minimum, possibly drastically overshooting it. Some program systems refer to this trick as 'damping'. It can be particularly important in cases of pseudo-symmetry, where much patience is required to get a structure away from a too-symmetric description.

The course of refinement can be directly monitored by the decreasing value of the minimization function or the various residual factors already defined. A more sensitive gauge of convergence is the ratio of each parameter shift to the corresponding parameter s.u. (these come directly from the matrix inversion in each cycle). For an acceptable convergence, all shift to s.u. ratios should be much smaller than unity; <0.01 is a reasonable aim. Parameters which are poorly determined by the data may give ratios which remain obstinately above this value, and in some cases they oscillate with positive and negative values in alternate cycles. This may indicate the inclusion of too many parameters (especially for poorly determined hydrogen atoms); addition of suitable constraints or restraints will often solve the problem.

Refinement usually involves progressively adding parameters until a satisfactory result is obtained. Each stage makes the model structure more complex (conversion from isotropic to anisotropic, incorporation of hydrogen atoms, modelling disorder, etc.). Whenever more parameters are added, the refinement has a greater degree of freedom, and a closer fit of observed and calculated data will usually be found. How do we know whether this improvement is significant or whether it is to be expected purely from the extra parameters? A particular case in point is the better fit which can often be obtained by lowering the space group symmetry: is it genuine? This question is addressed in Chapter 13, where some specific statistical tests are discussed. Here we note that a simple drop in R-indices does not necessarily mean an improved model. More reliable is a comparison of the final geometry s.u.s for the two models, which are likely to be a minimum for the most appropriate refinement. This is also a useful additional test for changes in the refinement weighting scheme.

12.7 Disorder

Disorder is a random (not systematic) variation in the detailed contents of the asymmetric unit of a crystal structure. For an ideal structure, all asymmetric units are exactly equivalent under the space group symmetry. Instantaneously, of course, this is not true, as each atom is undergoing vibration and these movements are not usually correlated throughout the structure, but this is covered by the ADPs, giving equivalence on a time-averaged basis. Large amplitudes of vibration are sometimes referred to as dynamic disorder.

If static disorder is truly random, then what X-ray diffraction sees is the average asymmetric unit (a systematic variation would lead to a larger asymmetric unit, possibly giving a structure/superstructure pseudo-symmetry problem.) This appears in the model structure as partially occupied atom sites.

Where the disordered atom sites are well resolved, refinement is likely to be much the same as for an ordered structure, except for the proper treatment of site occupancy factors for the disordered atoms; these need to be tied together with appropriate constraints.

Cases of disorder with atom sites which are closer than normal bonding distances are likely to require further constraints and/or restraints to assist the refinement. Each situation needs to be assessed individually, but a combination of suitable geometrical and ADP restraints, together with appropriate occupancy constraints, is a powerful tool to resolve even quite complex disorder.

It is not always immediately obvious whether large atomic displacement parameters are a genuine representation of high dynamic disorder, or whether they mask some kind of static disorder. Careful examination of electron density maps is essential. Low-temperature data collection is a tremendous advantage here: it reduces dynamic disorder and often makes it much easier to resolve and refine static disorder.

In the worst cases of disorder, usually affecting solvent molecules, individual atomic sites making geometrical sense cannot be resolved at all. Sometimes a collection of partial atoms must be refined as an approximate fit to the diffuse electron density. There are also methods available which include the diffuse electron density region in the model structure in some form other than as individual atoms, via Fourier transformation. Usually such regions are of no particular interest, but failure to deal with them in at least an approximate way reduces the precision of the structure as a whole.

12.8 Twinning

Crystal twinning is a problem that adds difficulties to the whole procedure of structure determination, from selection of crystals to the refinement and the interpretation of results. It is discussed in general in Chapter 18, to which reference should be made.

In order to handle twinning in refinement, the twin law needs to be found, describing the mathematical relationship between the two unit cell orientations; this is expressed as a 3×3 matrix relating the two sets of unit cell axis vectors and hence also the two sets of reflection indices. Provided some kind of starting model structure can be obtained and the twin law can be recognized, refinement is feasible. Each observed reflection intensity is the sum of two (or more) components and is fitted to

$$(F_c^2)_{\text{twin}} = k_1(F_c^2)_1 + k_2(F_c^2)_2 \tag{12.7}$$

where $k_2 = 1 - k_1$ (for the case of a two-part twin; multiple twins can also be handled by an extension of this). Thus the twin component fraction k_1 is a refined parameter and the twin law dictates which pairs of reflections must be combined together.

12.9   Absolute structure

As a first approximation, all diffraction patterns are centrosymmetric, even for non-centrosymmetric structures:

$$I(hkl) = I(\overline{hkl}) \tag{12.8}$$

— this is Friedel's law. It arises from the assumption, not completely true, that when X-rays are scattered by an atom, the phase of the scattered wave differs from that of the incident wave by a constant amount (180° or π radians). In fact, atoms of different elements introduce different phase shifts, and these are wavelength dependent. This effect is called anomalous scattering or anomalous dispersion, and is represented mathematically by adding a (usually small) imaginary component $i\Delta f''$ to each atomic scattering factor, thus making it complex instead of real (there is also a real anomalous scattering component $\Delta f'$). For a centrosymmetric structure, the anomalous scattering effects for reflections hkl and $\overline{hkl}$ cancel out, so that Friedel's law is rigorously maintained, but for non-centrosymmetric structures they do not, and $I(hkl) \neq I(\overline{hkl})$; the differences are usually small and demand careful measurement.

This difference means that the diffraction patterns for a particular non-centrosymmetric structure and for its inverse are not identical, and so it is possible, at least in principle, to determine the 'absolute structure'. In practice, with good quality data including Friedel pairs, this usually requires the presence of at least one silicon or heavier atom in the asymmetric unit with Mo $K\alpha$ radiation, and the presence of at least oxygen with Cu $K\alpha$ radiation.

The determination can be carried out by separate successive refinements of a structure and its inverse, which have exactly the same number of refined parameters, and comparison of the results. Alternatively, the structure can be refined as a racemic twin, one consisting of two components which are the inverse of each other, with twin law $(-1\,0\,0, 0-1\,0, 0\,0-1)$. The twin component parameter should refine to either zero or unity, depending on whether the current model structure is correct or inverted. This procedure has a number of advantages. It is computationally inexpensive, especially with F^2 refinement; it gives a numerical parameter whose s.u. provides an indication of the reliability of the result; and a result which is intermediate, with a small s.u., shows that the structure really is racemically twinned.

It is important always to carry out such a calculation for non-centrosymmetric structures, even if the absolute structure is itself either not of particular interest or believed to be known in advance. Refinement of the wrong handedness, particularly in a polar space group, can introduce systematic errors into the structure, particularly in the positions of heavy atoms, as the incorrect anomalous scattering effects are compensated by a shift of the atom along the polar direction; these errors can be much greater than the heavy atom positional s.u.s and give incorrect molecular geometry.

12.10 Other problems

Determining the correct space group is not always simple. Sometimes the distinction between two possible space groups giving the same systematic absences depends only on small differences in the model structure, such as the orientation of a single group of atoms. Sometimes the choice is between a structure which is disordered in a higher symmetry space group and one which is ordered (or less disordered) in a lower symmetry space group. The choice is not easy and there is no single rule for it. Attempted refinement with the lower symmetry often runs into problems of matrix singularity and lack of convergence because of the near-symmetry. Here, restraints (including shift-limiting restraints) and constraints may prove valuable, at least until the initial pseudo-symmetry has been broken. Attempts to refine a genuinely centrosymmetric structure in a non-centrosymmetric space group will certainly lead to singularity; even if this is overcome by some kind of trick, the resulting structure is likely to show severely distorted geometry because of huge correlations between pairs of parameters which are really related by symmetry. Removal of other symmetry elements will not lead to the same problems, but refinement in a space group of unnecessarily low symmetry should be avoided; look carefully at any structure which seems to have more than one molecule in the asymmetric unit, and use symmetry-checking programs as a matter of course.

Misassigned atom types will lead to incorrect ADPs, which can largely adjust to compensate. For example, a supposed nitrogen atom which is actually oxygen will have anomalously low ADPs; it may well also lead to residual electron density at this site, since the ADPs cannot entirely mop up the error. Ellipsoid plots and difference maps should not be neglected as tools for recognizing these and other problems.

Completely missing atoms may well have little effect on the refinement of other atoms, though R-indices will be somewhat high; they can really only be detected in (difference) electron density maps.

Overlooked disorder may manifest itself as unusual ADPs, particularly in the form of highly elongated ellipsoids which are attempting to cover two atom sites. Molecular geometry is also likely to be badly affected; watch out particularly for unusually short bonds to atoms with high ADPs.

In summary, the following should always be inspected carefully whenever problems are suspected, and also at the end of every refinement:

(a) R-indices and goodness of fit, as overall single-value indicators of the refinement fit;
(b) analysis of variance against indices, θ, and F_o^2, as an indication of suitable weighting scheme and inclusion of all appropriate parameters;
(c) list of 'outliers' for evidence of individual bad reflections;
(d) shift to s.u. ratios, as indicators of convergence;

(e) s.u.s of refined parameters as indicators of how well each parameter is determined by the data;
(f) data to parameter ratio, to check for over-parameterization;
(g) ADPs (preferably as principal axes and as ellipsoid plots) for reasonableness;
(h) molecular geometry for any unusual features and for correct treatment of H atoms;
(i) intermolecular contacts for any anomalies;
(j) absolute structure parameter for non-centrosymmetric structures, to check for correct handedness;
(k) difference map for any unaccounted features.

Exercises

12.1   What practical value might there be in using a weighting scheme
       (a) which is a direct function of $(\sin \theta)/\lambda$;
       (b) which is an inverse function of $(\sin \theta)/\lambda$?
12.2   How many refined parameters (excluding extinction) will there be for an ordered structure with two molecules of $C_6H_{12}O_6$ per unit cell, space group $P\bar{1}$:
       (a) for isotropic refinement without hydrogen atoms;
       (b) for anisotropic refinement with all hydrogen atoms riding so that they contribute no parameters themselves;
       (c) for anisotropic refinement of non-hydrogen atoms and free isotropic refinement of hydrogen atoms?
12.3   What are the constraints on coordinates and ADPs for an atom
       (a) on a twofold axis parallel to the c axis;
       (b) on a mirror plane in the 45° position between the a and b axes in a tetragonal space group?
12.4   Which (if any) are the axes with floating origins in the following space groups: (a) $P2_12_12_1$; (b) $P1$; (c) Cc; (d) $P2_1$?
12.5   A regular planar pentagon may be a useful rigid group for representing a poorly determined Cp ligand in a metal complex. Why should it not be used to represent a poorly determined tetrahydrofuran molecule, even if the oxygen and carbon atoms cannot be distinguished?
12.6   Suggest suitable restraints for
       (a) a CF_3 group attached to a benzene ring;
       (b) a cyclohexane solvent molecule in a chair conformation.
12.7   Why does a pseudo-symmetry problem often lead to very large calculated parameter shifts, requiring shift-limiting restraints?
12.8   Give examples of how a toluene solvent molecule may be disordered over an inversion centre.
12.9   Why does changing the signs of all atomic coordinates not produce the inverted structure in space group $P4_1$?
12.10 What is likely to happen in refinement if an atom is introduced into a completely incorrect position?
12.11 What is likely to be the effect on R-indices and on final geometry s.u.s of
       (a) leaving out weak reflections;
       (b) changing from isotropic to anisotropic displacement parameters for non-hydrogen atoms;
       (c) ignoring disordered solvent;
       (d) removing constraints on hydrogen atoms;
       (e) suppressing a few badly fitting reflections?

13

The derivation of results

13.1 Introduction

It is tempting to think that the final cycle of least-squares refinement marks the completion of a structure determination. There is, however, still much work to be done in analysing, interpreting and presenting the results, and sound experimental and computational procedures can be marred by failing to apply equally high standards to these final stages.

The *primary* numerical results of a structure determination are the parameters obtained in the least-squares refinement. Usually these consist of three positional coordinates and a number (often one, for isotropic, or six, for anisotropic models) of so-called 'temperature factors', 'thermal parameters' or 'displacement parameters' for each atom, together with a few other parameters for effects such as extinction and the overall scaling of the observed and calculated data sets. These parameters may have been refined as supposedly independent values, or there may be various constraints and/or restraints applied to their refinement (rigid groups, geometry restraints, riding hydrogen atoms, common displacement parameters for sets of atoms, etc.). The refinement supplies not only values for the parameters, but also an 'estimated standard deviation' (or e.s.d.) for each one (we shall consider later what these e.s.d. values mean). Recently the term 'standard uncertainty' (s.u.) has been introduced as a supposedly superior replacement for e.s.d. (although strictly they are not entirely equivalent), and we shall generally use the term s.u. here rather than e.s.d.

These *primary* results are usually not the main object of the structure determination experiment. Rather, we are interested in the molecular geometry and, possibly, in intermolecular interactions. Thus, we must derive *secondary* results: bond lengths, bond angles, torsion angles, and other distances and measures of geometry and conformation. (The term *secondary* here signifies only that these results are derived from the primary results, which are produced directly by the refinement process; there is no implication of secondary importance, as, indeed, the reverse is generally true.) If the geometrical results are to be of much value, we also need estimates of their reliability, in the form of an s.u. for each derived value. The calculation of both the secondary results and their s.u.s is, in principle, mathematically simple, but it is tedious, and usually it is hidden away in automatic 'black box' computer programs. We shall consider here what goes on inside these boxes, and explore some of the potential problems and sources of error and misconception.

As well as the atomic coordinates, from which molecular geometry is derived, the primary results include parameters describing atomic motions: the 'atomic displacement parameters' (ADPs). These are often ignored, but they do contain some useful and often important information. The analysis of these parameters is more difficult and open to much debate, and we shall consider some of their possible uses, including the interplay between positional and displacement parameters.

In the next chapter, we shall move on from the relatively straightforward *derivation* of results to the rather more vexed subject of their *interpretation*. Before both of these topics can be considered, however, it is necessary to lay a foundation with a brief treatment of the underlying mathematical and statistical ideas and language we will be using. Several general texts on structure determination include sections on this subject, and there are also more specialized treatments available [1–3].

13.2   Statistical background

13.2.1   Some basic mathematics and statistics

(a) Distributions
A frequency distribution or population is a set of observations or a function which describes the frequency with which different values are found for some measurable quantity, or the probability of finding these values for the quantity. Definitions like this are always difficult to express both generally and comprehensibly, and examples may help to illustrate the meaning. Thus, if we measure the heights of all the people in a particular group and tabulate or plot the number of people with height 150–155, 155–160, 160–165, ... cm, we shall have such a distribution. Other examples are: the distribution of ages of people living in a city; the population (number of molecules) of rotational energy levels for a diatomic molecule at a particular temperature (known as the Boltzmann distribution); the marks in a student examination. These distributions, drawn as graphs, will not all look the same shape! Examples are shown in Fig. 13.1.

If the values which can be taken by members of the distribution (values of the variable of the distribution) are only certain discrete ones (such as the quantized rotational energies and the examination marks mentioned above), we have a *discrete distribution*. If, on the other hand, the variables can take any value, possibly within a certain limited range (such as people's exact heights or ages), we have a *continuous distribution*.

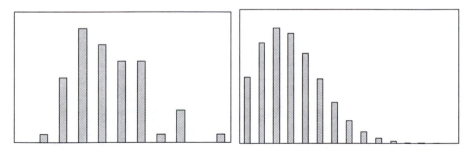

Fig. 13.1.   Some examples of distributions. *Left*: number of students scoring different numbers of marks out of 10 for a test. *Right*: population of rotational energy levels for molecules in the gas phase.

To describe or characterize a distribution, certain measures of its shape are commonly used. The two most important for our purposes are the *mean* and the *variance* (related to the *standard deviation*). For a discrete distribution, the mean is what we loosely call the 'average' value of the variable for the whole population:

$$\bar{x} = \frac{1}{N} \sum_{i=1}^{n} f_i x_i \tag{13.1}$$

where the N members of the population each take one of the n different values of x, and f_i is the number of members of the population having the value x_i. The variance of the distribution is defined as

$$\sigma^2 = \frac{1}{N} \sum_{i=1}^{n} f_i (x_i - \bar{x})^2 \tag{13.2}$$

and is a measure of the width or spread of the distribution over the different values of x. The variance is the square of the standard deviation σ.

For a continuous distribution [formally, $f(x)\,dx$ is the frequency/probability of a value between x and $x + dx$], the corresponding definitions are:

$$\bar{x} = \frac{\int_a^b x f(x)\,dx}{\int_a^b f(x)\,dx} \tag{13.3}$$

$$\text{and} \quad \sigma^2 = \frac{\int_a^b (x - \bar{x})^2 f(x)\,dx}{\int_a^b f(x)\,dx} \tag{13.4}$$

where a and b are the lower and upper limits of the variable x (which may be $-\infty$ and $+\infty$ in some cases). Note that if $f(x)$ is defined as a *probability* distribution, it is common to *normalize* it, such that

$$\int_a^b f(x)\,dx = 1 \tag{13.5}$$

(i.e. the total probability for all possible values of x is 1), so the denominators in the above expressions are not required.

Two particular distributions are of interest to us as crystallographers: the Poisson distribution and the Normal (or Gaussian) distribution.

(b) The Poisson distribution

This is applicable to problems involving a large number of experiments, in each of which there is a constant small chance of a particular event happening, or for large populations, to each member of which there is a constant small chance of a particular event happening. The distribution describes the probability of $0, 1, 2, \ldots, n$ such events occurring. It is appropriate, therefore, in assessing distinct events occurring in a continuum of time. Examples are the number of goals scored in a football match (though this ignores differences in skills of the teams!), deaths in a large-scale epidemic, or the number of times my telephone rings during the day. We shall not bother with the mathematical expression for the Poisson distribution. The important property is that its variance is equal to its mean.

In crystallography, the intensity output of an X-ray tube (in quanta per second) follows the Poisson distribution. Repeated measurement allows the mean and variance to be estimated, and they are found to be equal. It is on this basis that so-called 'counting statistics' are used to provide estimated standard deviations of diffractometer-measured intensities, for use in weighted least-squares structure refinement: more of this later.

(c) The Normal distribution
The mathematical expression for this very important distribution is

$$f(x) = \frac{1}{\sigma \sqrt{2\pi}} \exp\left(-\frac{(x - \bar{x})^2}{2\sigma^2}\right) \tag{13.6}$$

where $\bar{x}$ and σ^2 are the mean and variance, respectively. The distribution is symmetrical about its mean and the plotted shape is a familiar one in science. For a sufficiently large population (of people), heights will approximately follow the Normal distribution, as do frequently examination marks for large classes of students. The main characteristics of a Normal distribution are shown in Fig. 13.2.

In Fig. 13.2, 68.3% of a normal distribution lies within $\pm 1\sigma$ of the mean $\bar{x}$; the limits enclosing 90% and 95% are $\pm 1.65\sigma$ and $\pm 1.96\sigma$, while the interval $\pm 3\sigma$ encloses 99.7% of the total distribution.

The Normal distribution is particularly important for us because of an effect expressed by the *Central Limit Theorem*. Suppose we have a set of n independent variables x_i; each variable belongs to its own population with mean m_i and variance σ_i^2. Then, except in certain unusual circumstances, the function

$$y = \sum_{i=1}^{n} x_i \tag{13.7}$$

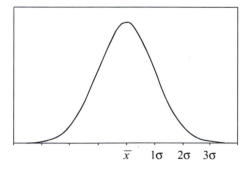

Fig. 13.2.   Characteristics of the Normal distribution.

has a distribution which, as n becomes very large, approaches a normal distribution with
mean and variance

$$m_y = \sum_{i=1}^{n} m_i \quad \text{and} \quad \sigma_y^2 = \sum_{i=1}^{n} \sigma_i^2 \tag{13.8}$$

whether the individual variables x have normal distributions or otherwise. It is generally
assumed that the experimental determination of the value of a particular quantity (such
as each of the parameters describing a crystal structure) is subject to a large number
of independent sources of small errors. Therefore, these quantities, if they were to be
measured repeatedly many times, would follow a normal distribution.

(d) Sampling a population

In order to estimate the mean age of the population of a city without having to collect
data for every person, we might try to select a random sample and calculate its mean
age, but how reliable would this be? In science, we are often concerned to measure some
mean quantity and to obtain an estimate of the reliability of our measurement.

Suppose we make n separate measurements of a quantity x. The measured values
$x_1 \cdots x_n$ are a sample from all the possible measurements we could make, which follow
some unknown distribution $f(x)$. For sufficiently large n, a consequence of the Central
Limit Theorem is that the mean $\bar{x}$ of our n sample values is normally distributed with
the same mean as the parent population (all possible measurements) and with variance
σ^2/n, where σ^2 is the variance of the parent population. By 'variance of the mean $\bar{x}$'
we understand the variance we would obtain by taking many such samples, calculating
the mean $\bar{x}$ for each separate sample, and then looking at the distribution (mean and
variance) of these individual sample means. It turns out that, if we have in fact taken
only one sample of measurements (as is usually the case), then the best estimate we
have for the population mean is our own sample mean $\bar{x}$, and the best estimate of the
population variance σ^2 is related to the variance of our own sample s^2 by

$$\sigma^2 = \frac{n}{n-1} s^2 \tag{13.9}$$

$$\text{with } s^2 = \frac{1}{n} \sum_{i=1}^{m} f_i (x_i - \bar{x})^2 \tag{13.10}$$

as usual, where there are m different values of x.
Our estimate of the variance of the mean $\bar{x}$ of the sample is thus

$$\sigma^2(\bar{x}) = \frac{1}{n(n-1)} \sum_{i=1}^{m} f_i (x_i - \bar{x})^2 \tag{13.11}$$

and this is a measure of the reliability or confidence with which we can use the sample
mean $\bar{x}$ as a true mean of the population, i.e. as the true value of the quantity we are
trying to measure. Note that the variance of the mean depends on the spread of individual
sample points x_i and inversely on the size of the sample n.

(e) Estimated standard deviations and standard uncertainties

This is all very well, but in crystallography we do not usually determine a structure several times in order to obtain mean values of the atomic parameters and estimates of the variance of these parameters! And yet from a *single* experiment we do obtain not only parameters but also estimated variances, more usually expressed as e.s.d.s or s.u.s. This is possible because of the fact that our experimentally measured data (diffraction intensities) greatly outnumber the parameters to be derived: the problem is said to be over-determined. In such cases, *estimated* standard deviations can be obtained from a single set of data, as we shall see.

13.2.2 Errors, precision and accuracy

Before we look in greater detail at the meaning, derivation and manipulation of estimated standard deviations and standard uncertainties in structural results, we must be clear about some other terms and how they are related.

Errors affecting experimental measurements are broadly of two types: random and systematic. *Random errors* are generally assumed to be normally distributed about a mean of zero, and they are unavoidable in experimental observations, although they may be minimized by careful measurement. *Systematic errors*, on the other hand, cannot be treated by any general statistical theory, and their presence may be quite unsuspected. As an example of the contrast, consider the measurement of a distance by means of a wooden metre rule. If the distance is measured by different people or repeatedly by one person, the separate measurements are likely to vary somewhat; this variation constitutes a random error in the measurement. If, however, the first 2 cm of the metre rule have been sawn off and this is not noticed, the measurements will be subject to a systematic error affecting all of them equally.

We must also distinguish between *precision* and *accuracy*. Precision concerns the question of how closely it is possible to make a particular measurement, or how confidently a derived result can be defined. It is related to the concept of *reproducibility*, and is measured by s.u.s. Accuracy, on the other hand, concerns the question of how well our measurement or result agrees with the true value we aim to measure or derive. We may achieve high precision with our faulty metre rule (it could be even higher if we used sophisticated optical equipment instead), but the accuracy is poor!

Truly random errors affect the precision but not the accuracy of measurements and results. Depending on their exact nature, systematic errors may or may not affect precision, but they almost certainly affect accuracy (Fig. 13.3). Thus, high precision is not of itself an indication of a 'good' result.

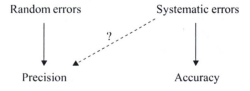

Fig. 13.3. The effect of random and systematic errors on precision and accuracy.

Random errors are treated by statistics: standard deviations and estimated standard deviations are direct measures of precision. Systematic errors must be dealt with by proper corrections, allowed for in our results, or avoided in the first place. They may arise in the measurement of the data (e.g. effects of absorption or a badly aligned diffractometer) or in the models and methods used in structure determination (e.g. incorrect atomic scattering factors, inappropriate atomic displacement parameters, wrong space group symmetry).

13.2.3 Estimated standard deviations/standard uncertainties in crystallographic results [4]

To recap so far, for each refined or derived parameter, there is a particular value and a corresponding s.u. The value obtained for the parameter is our best estimate of the true value. The s.u. is a measure of the precision or statistical reliability of this value; it is our best estimate of the variation we would expect to find for this parameter if we were to repeat the whole experiment many times.

(a) Standard uncertainty (s.u.) values for refined parameters
In structure refinement by least squares, a number of parameters are determined from a larger number of observed data. The quantity minimized is

$$\sum_{i=1}^{N} w_i \Delta_i^2 \tag{13.12}$$

where Δ_i is usually either $|F_o|_i - |F_c|_i$ or $F_{o,i}^2 - F_{c,i}^2$ and each of the N reflections has a weight w_i. The s.u. values of the refined parameters depend on (i) the minimized function; (ii) the numbers of data and parameters; and (iii) the diagonal elements of the inverse least-squares matrix $\mathbf{A}^{-1}$

$$\sigma(p_j) = \left((\mathbf{A}^{-1})_{jj} \frac{\sum_{i=1}^{N} w_i \Delta_i^2}{N - P} \right)^{1/2} \tag{13.13}$$

where p_j is the jth of the P parameters.

Note that low s.u. values (high precision) are achieved with a combination of good agreement between observed and calculated data (small numerator) and a large excess of data over parameters (large denominator).

(b) Correlation and covariance [5]
The parameters describing a crystal structure are not independent. When we derive further results from a combination of several parameters (such as calculating a bond length from the six coordinates of two atoms), it is important to recognize this interrelationship to calculate the correct s.u.s for the secondary results.

When variables are not statistically independent, they are said to be *correlated*. Just as individual variables have *variances*, so correlated variables have *covariances*.

For a discrete distribution of two correlated variables x and y, the covariance is defined as

$$\text{cov}(x, y) = \frac{1}{N} \sum_{i=1}^{n} f_i (x_i - \bar{x})(y_i - \bar{y}) \tag{13.14}$$

which should be compared with the variances

$$\sigma^2(x) = \frac{1}{N} \sum_{i=1}^{n} f_i (x_i - \bar{x})^2 \quad \text{and} \quad \sigma^2(y) = \frac{1}{N} \sum_{i=1}^{n} f_i (y_i - \bar{y})^2. \tag{13.15}$$

Thus, $\text{cov}(x, x) = \sigma^2(x)$ by definition.

For a continuous distribution, similarly

$$\text{cov}(x, y) = \frac{\int_a^b \int_c^d (x - \bar{x})(y - \bar{y}) f(x, y) \, dx \, dy}{\int_a^b \int_c^d f(x, y) \, dx \, dy} \tag{13.16}$$

where c and d are the lower and upper limits of the variable y.

In both cases, the *correlation coefficient* of x and y is

$$r(x, y) = \frac{\text{cov}(x, y)}{\sigma(x)\sigma(y)} \tag{13.17}$$

and this must lie in the range ± 1. A correlation coefficient of exactly $+1$ or -1 means that x and y are perfectly correlated—each is an exact linear function of the other, and only one variable is required to describe them both. If x and y are completely independent, their covariance and correlation coefficient are both zero, although the converse is not necessarily true (a covariance of zero does not have to mean independent variables).

Covariances (and hence correlation coefficients) of all pairs of refined parameters for a crystal structure are obtained together with the variances from the inverse matrix:

$$\text{cov}(p_j, p_k) = (\mathbf{A}^{-1})_{jk} \frac{\sum_{i=1}^{N} w_i \Delta_i^2}{N - P}. \tag{13.18}$$

Once we have all the variances and covariances of a set of quantities (such as the refined atomic parameters), we can calculate the variances and covariances of any functions of these quantities (such as molecular geometry parameters).

For a function $f(x_1, x_2, x_3, \ldots, x_n)$

$$\sigma^2(f) = \sum_{i,j=1}^{n} \frac{\partial f}{\partial x_i} \cdot \frac{\partial f}{\partial x_j} \cdot \text{cov}(x_i, x_j) \tag{13.19}$$

where $\text{cov}(x_i, x_i)$ is the same as $\sigma^2(x_i)$, as we saw before.

For two functions f_1 and f_2

$$\text{cov}(f_1, f_2) = \sum_{i,j=1}^{n} \frac{\partial f_1}{\partial x_i} \cdot \frac{\partial f_2}{\partial x_j} \cdot \text{cov}(x_i, x_j) \qquad (13.20)$$

but this is not often needed.

Note that if the variables x are all independent, the covariances are zero except for the variance terms themselves, so in such a simple case

$$\sigma^2(f) = \sum_{i=1}^{n} \left(\frac{\partial f}{\partial x_i} \right)^2 \sigma^2(x_i) \qquad (13.21)$$

and thus the variance of f is just a weighted sum of the variances of the individual independent variables.

The full variance–covariance matrix following the final least-squares refinement must, therefore, be used in calculating molecular geometry s.u. values. Calculation using the coordinate s.u. values alone, with neglect of correlation effects between atoms, does not give the correct geometry s.u.s: it is equivalent to using the last equation above instead of the previous one, and potentially important terms are missing. This is particularly evident when calculating the bond length between two atoms related by a symmetry element, because the coordinates of these atoms are completely correlated.

Such calculations are normally performed automatically together with the least-squares refinement. Proper calculation in a separate step later would require that the refinement program output the full variance–covariance matrix.

13.3   Analysis of the agreement between observed and calculated data

Before we are able to calculate the desired geometrical information from the refined atomic parameters, refinement must reach convergence. But is this convergence the best that can be achieved for our structure? At several stages during a typical structure determination, refinement converges, but the introduction of more parameters (a change in the model being refined) allows further refinement to take place, giving convergence again, this time (we hope) to an even better agreement with the observed data. Such developments of the model are, for example, the replacement of isotropic by anisotropic atomic displacement parameters, and the inclusion of hydrogen atoms. Other changes which can be made are the refinement of parameters for effects such as extinction or absorption, the addition of models of disorder, and changes to the weighting scheme. Any change in the model will produce a different set of refined parameters. How do we assess the agreement with the observed data and choose the 'best' model?

13.3.1   Observed and calculated data

Overall 'residuals' (single-value measures of the agreement) commonly used and quoted are:

$$R = \frac{\sum_i |\Delta|_i}{\sum_i |F_o|_i}$$

$$wR = \left(\frac{\sum_i w_i \Delta_i^2}{\sum_i w_i F_{o,i}^2}\right)^{1/2} \tag{13.22}$$

$$S = \left[\frac{\sum_i \{\Delta_i/\sigma(F_o)_i\}^2}{N - P}\right]^{1/2}$$

as discussed in Chapter 12. R is long established as *the* traditional 'R-factor', accorded a general reverence far in excess of its real significance. The 'generalized R-factor', called wR here in accordance with current *Acta Crystallographica* usage, is variously referred to as RG, R' and R_w as well. S is the so-called 'goodness of fit', which should have a value of 1 if the model is a true representation of the X-ray scattering power of the structure, the $\sigma(F_o)$ values are correct and on an absolute scale, and errors are random only: this is almost never achieved in practice! Note that similar residuals can be defined with F^2 replacing F if the refinement is based on F^2 values.

All of these residuals can be manipulated and massaged in various ways to produce an apparently better fit to the data. Both R and wR can be reduced dramatically by omitting some reflections, especially the weak ones and a few which give particularly poor agreement between $|F_o|$ and $|F_c|$ (perhaps on grounds such as 'they are strongly affected by extinction'). Altering the weighting scheme changes all the residuals, and S can artificially be made equal to 1 simply by rescaling all the $\sigma(F_o)$ values by the necessary factor. Note that both wR and S are closely related to the least-squares minimization function $\sum_i w_i \Delta_i^2$.

A much better assessment of the fit of observed and calculated data comes from an analysis of the discrepancies in terms of particular groups of reflections rather than just for the data set as a whole. Any of the residuals (and others related to them) can be calculated and tabulated, for example, for reflections in different index parity classes, different ranges of $(\sin\theta)/\lambda$, different ranges of $|F_o|$ or F_o^2, different values of each of the indices h, k, l, or any other desired groupings. Trends in the values of residuals across different groups (especially different ranges of Bragg angle and of structure factor amplitude) reveal the presence of systematic errors in the model (e.g. neglect of hydrogen atoms or of extinction effects) or imperfections in the weighting scheme. Indeed, empirical adjustments to the weighting scheme can be made on the basis of such an analysis; the weights thus obtained are supposed to reflect not only uncertainties in the data but also shortcomings of the structural model. Simpler versions of empirical adjustments are included in most refinement programs, especially the addition of terms depending on F^2 to the weighting scheme.

Another method of assessing the data fit is through a 'normal probability plot' analysis. Such plots are useful, not only here but in any comparison of two sets of quantities

(e.g. two independently measured data sets for the same structure, or two sets of parameters refined from them). For data sets, we define a weighted deviation for each pair of
observations as

$$\delta_i = \frac{(|F_o|_i - |F_c|_i)}{\sigma(F_o)_i} \tag{13.23}$$

for observed and calculated data, or

$$\delta_i = \frac{F_{1,i} - F_{2,i}}{[\sigma^2(F_{1,i}) + \sigma^2(F_{2,i})]^{1/2}} \tag{13.24}$$

for the comparison of two independent sets of measured data. The deviations δ_i are
then sorted into order of increasing value from the most negative to the most positive,
and the sorted list is compared with the values which would be expected for a normal
(or any other desired) distribution; these values may be obtained from standard tables or
by calculation. The observed and expected δ_i are then plotted against each other. If the
actual distribution of the δ_i values matches the expected distribution, with no systematic
errors present, the plot will be a straight line of gradient $+1$, passing through the point
(0,0). Usually the plot is made against an expected normal distribution with zero mean
and unit variance, corresponding to random errors which are faithfully represented by
the $\sigma^2(F)$ values. An example is shown in Fig. 13.4.

Thus, a linear plot indicates that the errors are, indeed, essentially random; systematic
errors produce non-linearity. A gradient differing from unity indicates that the $\sigma^2(F)$
values are either underestimated (gradient > 1) or overestimated (gradient < 1). Note
that this gradient and the goodness of fit S are closely related. If the intercept of the plot
is not zero, the two sets of data are not correctly scaled relative to each other.

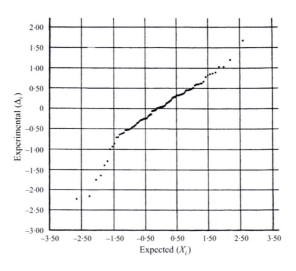

Fig. 13.4.   A normal probability plot.

The use of normal probability plots for comparing sets of structural parameters is discussed in the next chapter.

Such detailed analyses of the data fit also help to reveal 'outliers'—individual reflections which give much worse agreement than the rest. These can then be investigated (perhaps there was a problem during data collection?). Unrecognized outliers can have a considerable effect on overall residuals, and may significantly affect the final refined parameters.

13.3.2   Significance testing

Increasing the number of refined parameters will always (given suitable weights) reduce the residuals and produce a better fit to the observed data. It does not follow that any such reduction is significant and meaningful.

There are standard tests for assessing statistically the improvement in fit when the model is changed. The ratio of the wR residuals is calculated

$$\Re = \frac{wR(1)}{wR(2)} \tag{13.25}$$

and compared with tabulated values. The tables are constructed for different 'dimensions' (the difference in number of parameters for models 1 and 2) and 'degrees of freedom' ($N - P$ for model 2) and according to various 'levels of significance' α. Thus, if the value of $\Re$ is greater than a tabulated value appropriate to the degrees of freedom and dimension of the test, this means that the probability that this apparent improvement could arise by chance from two equally good models is less than α; model 2 is said to be better than model 1 at the α significance level (α is often expressed as a percentage). So a significance level of $\alpha = 0.01$ (1%), for example, indicates that there is a 1% risk of accepting the second model as better when in fact it is not.

Interpolation is often necessary between tabulated values, because the available tables do not cover exactly the dimension and degrees of freedom required. In practice, it is rare for these tests to indicate that improvement is not significant, and there are some doubts as to their true statistical validity. Alternative tests have, therefore, been proposed.

Note that the s.u.s of the refined parameters do not necessarily decrease when the residuals decrease. Although they do depend on the minimization function $\sum_i w_i \Delta_i^2$, they also depend *inversely* on the excess of data to parameters, $N - P$. A very simple assessment of the significance of the improvement of a model on introducing extra parameters is, then, to see whether the parameter s.u. values are reduced.

13.4   Geometry

So now we have a converged refinement with which we are satisfied. The primary results include three coordinates for each atom. The secondary results, generally of greater interest, are parameters describing the molecular geometry.

13.4.1   Bond lengths, bond angles and torsion angles

Several standard crystallographic texts give equations for the calculation of molecular geometry parameters from atomic coordinates and unit cell dimensions. These appear to be very different from one another, because they are variously expressed in trigonometric or vector notation, and some involve intermediate conversion from unit cell axes to arbitrary orthogonal axes.

For a triclinic structure, the bond length (or, generally, the distance) l between two atoms is given by

$$l^2 = (a\Delta x)^2 + (b\Delta y)^2 + (c\Delta z)^2 - 2bc\cos\alpha\,\Delta y\Delta z$$

$$- 2ac\cos\beta\,\Delta x\Delta z - 2ab\cos\gamma\,\Delta x\Delta y \tag{13.26}$$

and a bond angle θ for three atoms A–B–C is given by the 'cosine rule'

$$\cos\theta = \frac{l_{BA}^2 + l_{BC}^2 - l_{AC}^2}{2l_{BA}l_{BC}}. \tag{13.27}$$

Torsion angles measure the conformational twist about a series of four atoms bonded together in sequence in a chain A–B–C–D. The torsion angle is defined as the rotation about the B–C bond which is required to bring B–A into coincidence with C–D when viewed from B to C. The generally accepted sign convention is that a positive torsion angle corresponds to a clockwise rotation. Note that (i) the torsion angle D–C–B–A is identical in magnitude *and* sign to the torsion angle A–B–C–D, so there is no ambiguity in the description of the angle; and (ii) the torsion angles for equivalent sets of atoms in a pair of enantiomers have equal magnitudes but opposite signs, so that all torsion angles change sign if a structure is inverted. Formulae for the calculation of torsion angles are given in detailed crystallography texts.

Although the occasional distance or angle can be calculated by hand (for example, where a non-bonded distance is required, but is not listed automatically by the refinement program because it is too long), these derivations are tedious, and are best left to automatic computer programs. Even more to the point, the correct calculation of s.u.s for the molecular geometry parameters requires inclusion of covariance terms, because atomic coordinates are not uncorrelated; the necessary covariances, produced automatically by a full-matrix least-squares refinement (note: refinements not based on a full matrix do not give all the covariances, and also tend to underestimate variances), are not normally preserved and printed after the refinement, so only approximate s.u.s can be hand-calculated, on the assumption of uncorrelated atoms, from the atomic coordinate s.u. values. The approximation will be a particularly poor one when symmetry-equivalent atoms are involved, e.g. for a bond across an inversion centre or an angle at an atom on a mirror plane. Formulae are given for uncorrelated parameters in some standard references.

Note that any parameter which is varied in the least-squares refinement will have an associated s.u., and any parameter which is held fixed will not. Usually, the three coordinates and six anisotropic U^{ij} (or one isotropic U) for each atom are refined, and each has an s.u. Symmetry may, however, require that some parameters are fixed, because atoms lie on rotation axes, mirror planes or inversion centres; in this case, the s.u. of such a fixed parameter must be zero. This has an effect on the s.u.s of bond lengths and other geometry involving these atoms, which will tend to be smaller than they would be for refined parameters. If a coordinate of an atom has been fixed in order to define a floating origin in a space group with a polar axis (better methods are used in most modern programs), the effect will be to produce artificially better precision for the geometry around this atom.

Parameters which are equal by symmetry must have equal s.u.s. This applies both to the primary refined parameters (for example, atoms in certain special positions in high-symmetry space groups have two or more equal coordinates and relationships among some of the U^{ij} components), and also to the geometrical parameters calculated from them. A good test of the correctness of the calculation of geometry s.u.s by a program is to compare the bond lengths and their s.u.s for atoms in special positions in trigonal and hexagonal space groups!

If a bond length (or other geometrical feature) has been *constrained* during refinement, the s.u. of this bond length must necessarily be zero, even though the two atoms concerned will, in general, have non-zero s.u.s for their coordinates; this is a consequence of correlation: the covariance terms exactly cancel the variance terms in calculating the bond length s.u. from the coordinate s.u.s. A good example of such a situation is the 'riding model' for refinement of hydrogen atoms, where the C–H bond is held constant in length and direction during refinement. The C and H atoms have the same s.u.s for their coordinates (because they are completely correlated), and the C–H bond length has a zero s.u. By contrast, *restrained* bond lengths do have an s.u., because the restraint is treated as an extra observation and the two atoms are in fact refined normally. It is instructive to compare the calculated bond length and its s.u. with the imposed restraint value and its weight, to see how valid the restraint is in the light of the diffraction data.

For a group of atoms refined as a 'rigid group', all internal geometrical parameters will have zero s.u.s. The parameters actually refined are the three coordinates for some defined point in the group (usually one atom or the centroid) and three rotations for the group as a whole. Thus, different atoms in the group should have different coordinate s.u.s (this is not the case with some refinement programs, which do not calculate these s.u.s correctly), and, once again, the effects of correlation and covariance are such as to give the required zero s.u.s for the geometry of the group.

It is often overlooked that molecular geometry depends not only on the atomic coordinates but also on the unit cell parameters, and they too are subject to uncertainties. Some refinement programs make no allowance for the uncertainties in the cell parameters, and the results can be ridiculous, especially for the geometry around heavy atoms. These usually have very low coordinate s.u. values, so bond lengths and angles calculated without regard to cell parameter uncertainties may have s.u.s proportionately very much smaller than the cell edge s.u.s! If explicit treatment of this effect is not included in your geometry calculation program, a simple hand adjustment can be made, by increasing s.u.s by an amount depending on the ratio of the cell edges to their s.u.s $[\sigma(a)/a, \sigma(b)/b$ and $\sigma(c)/c]$; this usually affects only the heaviest atoms in a structure of contrasting atomic scattering powers.

13.4.2 Least-squares planes and dihedral angles

It is sometimes desirable to assess whether a number of atoms are all in one plane and, if not, how much they deviate from coplanarity; this is particularly the case for cyclic groups of atoms and for selected atoms coordinating a central metal atom. When more than one exact or approximate plane can be defined in a structure, the angles between pairs of planes may also be of interest.

The usual method of assessing planarity of a group of atoms is to fit an exact plane to the atomic positions by a least-squares calculation; the plane is chosen so as to minimize $\sum_{i=1}^{n} w_i \Delta_i^2$, where Δ_i is the perpendicular distance of the ith atom from the plane, there are n atoms to be fitted, and each has relative weight w_i in the calculation. There are various methods of performing the calculation, which can also be expressed as a determination of one of the principal axes of inertia for the group of atoms. In the calculation, the weights used for the atoms should be proportional to $1/\sigma^2$, where σ^2 is the variance for the atomic position in the direction perpendicular to the required plane. As a reasonable approximation, an overall average positional σ^2 may be used for each atom, but even this is often not done and unit weights are used instead. A very crude approximate scheme weights atoms proportionally to their atomic numbers or atomic masses, since the heaviest atoms usually have the smallest positional s.u. values.

Calculation of a least-squares plane also provides a 'root-mean-square deviation' of the atoms from the plane

$$\text{r.m.s.}\Delta = \left(\frac{\sum_{i=1}^{n} w_i \Delta_i^2}{n} \right)^{1/2} \tag{13.28}$$

and this quantity may be used to assess whether the deviations from planarity are significant. A standard statistical test (the χ^2 test) can be applied, but it is rare for any set of more than three atoms to be judged truly planar by this test (except for groups which are strictly planar by symmetry, such as four atoms related in pairs by an inversion centre). It is common to quote the deviations of individual atoms from the plane; these atoms may be among those used to define the plane itself, or may be other atoms. The calculation (and even the definition) of an s.u. for such a deviation is not obvious, and various accounts have been given. Generally, these involve considerable approximations and the neglect of correlation effects.

Deviations of atoms from a least-squares plane are a much more sensitive, and hence a better, estimate of whether the coordination about an atom is essentially planar than is the sum of the bond angles at that atom. This sum will be quite close to $360°$ even for a markedly pyramidal three-coordinate atom or for a square-planar coordination with significant tetrahedral distortion.

The terms 'least-squares plane' and 'mean plane' are used synonymously by most crystallographers, although some authors have distinguished between them, giving them different definitions.

The angle between two planes is sometimes called a dihedral angle, though this term may also be used to mean the same as torsion angle, so some care is needed. We must also be aware of an ambiguity in defining the interplanar angle. The correct definition is the angle between the normals to the two planes, but where two lines cross a choice can be made between two possible angles, whose sum is 180°. Where the two planes concerned have two atoms in common, the dihedral angle represents a fold about the line joining these two atoms (a 'hinge' or 'flap' angle), and it seems sensible to choose the angle enclosed by the two hinged planes (so that the angle would be 0° for a closed hinge and 180° for a fully opened hinge), but the choice of angle is less obvious in some other situations.

13.4.3 Conformations of rings and other molecular features

It is in describing molecular features such as coordination, planes and ring conformations that we move from unambiguous description to interpretation. The conformation of rings can be described in many ways [6]. The most common quantities used to describe ring conformations are torsion angles, atomic deviations from least-squares planes, and angles between these planes, and on the basis of such measures, rings are generally classified by such terms as chair, boat, twist, envelope, etc. Ring conformations can also be analysed in terms of linear combinations of normal atomic displacements according to irreducible representations of the D_{nh} point group symmetry appropriate to a regular planar n-membered ring.

The shapes of coordination polyhedra around a central atom can also be difficult to describe, and we frequently see simple expressions such as 'distorted tetrahedral' or 'approximately octahedral', which may refer to extremely unsymmetrical arrangements! Attempts to quantify these descriptions have included definitions of 'twist angles' and other measures of the degree of distortion from regular coordination shapes.

13.4.4 Hydrogen atoms and hydrogen bonding

The distance between two atoms, together with its s.u., can be calculated regardless of whether the two atoms are considered to be bonded to each other. Distances between atoms in adjacent molecules may indicate significant intermolecular interactions, if they are shorter than some 'expected' or standard value (such as the sum of van der Waals radii for the atoms concerned). Short contacts involving a hydrogen atom and an electronegative atom are often examined as potential candidates for hydrogen bonding. There are, however, some pitfalls to be avoided here.

First, hydrogen atoms are not very precisely located by X-ray diffraction, because of their low electron density. Thus, freely refined hydrogen atoms will have larger positional s.u.s than other atoms. Some computer programs list bond lengths with s.u.s, but non-bonded distances without any estimate of precision. The relatively low precision of these distances should not be overlooked in interpreting the distances themselves. Weak hydrogen bonding is sometimes postulated when the experimental precision simply does not support it.

Second, hydrogen atom positions determined by X-ray diffraction do not correspond to true nuclear positions, because the electron density is significantly shifted towards the atom to which the hydrogen atom is covalently bonded. Thus, typical bond lengths for freely refined atoms are around 0.95 Å for C–H and under 0.90 Å for N–H and O–H, whereas true internuclear distances, obtained by spectroscopic methods for gas-phase molecules or by neutron diffraction, are over 0.1 Å longer. In hydrogen bonding, the hydrogen atom lies roughly between its covalently bonded atom and the electronegative atom in an X–H $\cdots$ Z arrangement, so a significant shortening error in the X–H bond length means an incorrectly long H $\cdots$ Z distance. This is another reason why these distances should be interpreted with caution.

Third, hydrogen atoms are constrained (or restrained) in many structure determinations, and their positions are, therefore, to a large extent dictated by preconceived ideas. Hydrogen bonding of any significance is likely, however, to perturb hydrogen atoms from 'expected' positions.

For these reasons, the X $\cdots$ Z distance may often be a better (or at least a safer) indication of hydrogen bonding. In any case, possible hydrogen bonding which does not fit in with widely recognized patterns should be examined very carefully before it is presented to the public [7]!

13.5 Thermal motion

Although the major interest in a structure determination usually centres on the geometry derived from the atomic positions, the primary results also include the so-called 'thermal parameters'. It has been suggested that these describe not only the time-averaged temperature-dependent movement of the atoms about their mean equilibrium positions (*dynamic disorder*), but also their random distribution over different sets of equilibrium positions from one unit cell to another, representing a deviation from perfect periodicity in the crystal (*static disorder*) which is not great enough to be resolved into distinct alternative sites, and so they should rather be called 'atomic displacement parameters'. Much of the literature on this subject is complicated to read, being expressed in unfamiliar notation; indeed, it has been pointed out that some standard textbooks give a misleading interpretation of these parameters. A refreshingly readable account has been written for a general chemical audience, and is strongly recommended [8].

Interpretation and analysis of displacement parameters is not often undertaken. One reason is that various systematic errors in the data, inappropriate refinement weights and poor aspects of the structural model all tend to affect these parameters, whereas the atomic positions are much less perturbed (fortunately!). Thus, the 'anisotropic temperature factors' of a structure are often regarded as a sort of error dustbin, and their physical significance is questionable unless the experimental work is of good quality.

13.5.1 β, B and U parameters

It is unfortunate that atomic displacements are described by a variety of different parameters, all of which are mathematically related. Thus, for an isotropic model, a single parameter is used, but this may be called B or U. These are related by

$$f'(\theta) = f(\theta) \exp(-B \sin^2\theta/\lambda^2) = f(\theta) \exp(-8\pi^2 U \sin^2\theta/\lambda^2) \qquad (13.29)$$

where $f(\theta)$ is the scattering factor for a stationary atom and $f'(\theta)$ the scattering factor for the vibrating atom. B and U both have units of $Å^2$ and U represents a mean-square amplitude of vibration.

For an anisotropic model, six parameters are used, and the exponent $(-B\sin^2\theta/\lambda^2)$ becomes variously

$$- (\beta^{11}h^2 + \beta^{22}k^2 + \beta^{33}l^2 + 2\beta^{23}kl + 2\beta^{13}hl + 2\beta^{12}hk)$$

$$\text{or}\quad -\tfrac{1}{4}(B^{11}h^2a^{*2} + B^{22}k^2b^{*2} + B^{33}l^2c^{*2} + 2B^{23}klb^*c^*$$
$$+ 2B^{13}hla^*c^* + 2B^{12}hka^*b^*) \qquad (13.30)$$

$$\text{or}\quad -2\pi^2(U^{11}h^2a^{*2} + U^{22}k^2b^{*2} + U^{33}l^2c^{*2} + 2U^{23}klb^*c^*$$
$$+ 2U^{13}hla^*c^* + 2U^{12}hka^*b^*).$$

The first form is most compact, but the six β terms are not directly comparable (the factor 2 in the three cross-terms is sometimes omitted, adding yet more confusion to the possible definitions); the second form is equivalent to the isotropic B, and the third to the isotropic U expression. B and U forms are very easily interconverted; we shall use the U forms here.

These parameters are often represented graphically as 'thermal ellipsoids'. Note that this is possible only if certain inequality relationships among the six parameters are satisfied; otherwise they are said to be 'non-positive definite' and the corresponding ellipsoid does not have three real principal axes. Such a situation may indicate a real problem in the structural model (e.g. a disordered atom), or it may just be due to imprecise U^{ij} parameters (high s.u.s), in which case the anisotropic model for this atom is perhaps not justified.

13.5.2   'The equivalent isotropic displacement parameter'

Tables of anisotropic displacement parameters are very unlikely to be published in most chemical journals, and their significance is difficult to assess at a glance. For a simple assessment of the atomic motions, it is convenient to calculate an equivalent isotropic parameter for each atom, U_{eq}, and some journals prefer these to be included in tables of atomic coordinates. Different definitions of U_{eq} abound, and some of them seem to be inappropriate [9]. Essentially, one version of the equivalent isotropic parameter is that corresponding to a sphere of volume equal to the ellipsoid representing, on the same probability scale, the anisotropic parameters. The definition '$U_{eq} = \tfrac{1}{3}$ (trace of the orthogonalized U^{ij} matrix)' is a commonly used one, but its meaning is, perhaps, not entirely clear! It can be expressed mathematically as (among other equivalent forms)

$$U_{eq} = \frac{1}{3}\sum_{i=1}^{3}\sum_{j=1}^{3} U^{ij}a_i^*a_j^*\mathbf{a}_i\cdot\mathbf{a}_j \qquad (13.31)$$

where the direct and reciprocal cell parameter terms have the effect of converting the U^{ij} parameters into a form expressed on orthogonal rather than crystal axes.

Acta Crystallographica Section C introduced a requirement for U_{eq} values in coordinate tables in 1980. Corresponding s.u. values were also required, but this demand was dropped later. A simple calculation of the s.u. for U_{eq} can be made from the s.u.s of the U^{ij} parameters, but it has been shown that a proper inclusion of covariance terms (correlations among the U^{ij} values), which are not always available after the refinement is complete, gives lower s.u. values, so the simply calculated values are of dubious worth.

13.5.3   Models of thermal motion and geometrical corrections: rigid body motion

It is well known that one effect of thermal vibration is to produce an apparent shrinkage in molecular dimensions. Analysis of this effect and correction for it are possible only in certain cases.

If a molecule has only small internal vibrations (both bond stretching and angle deformations) compared with its movement as a whole about its mean position in a crystal structure, then it can be treated approximately as a rigid body. In this case, the movements of the individual atoms are not independent and so the U^{ij} parameters of the atoms must be consistent with the overall molecular motion. This motion can be described by a combination of three tensors (3×3 matrices): the overall *translation* (oscillation backwards and forwards in three dimensions), represented by the six independent components of a symmetric tensor T (analogous to the anisotropic U tensor for an individual atom); *libration* (rotary oscillation), represented also by a symmetric tensor L; and *screw* motion, represented by an unsymmetrical tensor S. This third contribution is necessary to describe the complete motion of a molecule which does not lie on a centre of symmetry in the crystal structure, because there is then correlation between translational and librational motion, such that the librational axes do not all intersect at a single point. The tensor S in fact has only eight independent components, because the three diagonal terms are not all independent, so the whole molecular motion can be described by 20 parameters. Except for very small molecules (and some particular geometrical shapes), the six U^{ij} values for each atom provide more than enough data for a least-squares refinement to determine these 20 parameters, and the agreement between observed and calculated U^{ij} values gives a measure of the usefulness of the rigid body model.

From the rigid body parameters, corrections can be calculated for bond lengths within the molecule; these depend only on the librational tensor components.

Although many molecules cannot be regarded as even approximately rigid, it may be possible to treat certain groups of atoms within them as rigid bodies, and make corrections within those groups.

It is possible to test whether a molecule, or part of a molecule, might be regarded as a rigid body [10]. If a pair of atoms (whether bonded together directly or not) behaves as part of a rigid group, then they must remain at a fixed distance apart during their concerted motion. In this case, the components of their individual anisotropic vibrations along the line joining them must be equal. Thus a 'rigid atom-pair' test computes these components of anisotropic motion:

$$\langle U^2 \rangle = U^{11}d_1^2 + U^{22}d_2^2 + U^{33}d_3^2 + 2U^{23}d_2d_3 + 2U^{13}d_1d_3 + 2U^{12}d_1d_2 \qquad (13.32)$$

where $\langle U^2 \rangle$ is the mean-square amplitude of vibration along a line which has direction cosines d_1, d_2, d_3 referred to reciprocal cell axes. Equality or near-equality of the $\langle U^2 \rangle$

values for the two atoms is a necessary (but not sufficient) condition for rigidity. This can be used as a test for rigid bonds and for rigid bodies (the test must work for every pair of atoms in the group being tested). It can also be used as the basis of a restraint on U^{ij} values in structure refinement.

13.5.4 Temperature and atomic displacement parameters

Although the U^{ij} parameters of the atoms probably do not describe only thermal vibration effects, as noted above, they are usually strongly temperature dependent, and they can be drastically reduced by carrying out data collection at a lower temperature. With reliable low-temperature apparatus now available for X-ray diffractometers, this approach is strongly to be recommended. Low-temperature data usually give greater precision in atomic positions, more reliable molecular geometry, and an opportunity to assess and distinguish between dynamic and static disorder: the former will be reduced at lower temperature, the latter will probably not. Although we are concerned in this chapter with the analysis of results, we should bear in mind that this analysis can be greatly helped by an improvement in the experimental measurements!

References

[1] E. Prince, *Mathematical Techniques in Crystallography and Materials Science*, Springer, New York, 1982.
[2] W. C. Hamilton, *Statistics in Physical Science*, Ronald Press, New York, 1964.
[3] K. Huml, *Computing in Crystallography*, ed. R. Diamond, S. Ramaseshan and K. Venkatesan, Indian Academy of Sciences, Bangalore, 1980, Chapter 12.
[4] D. Schwarzenbach, S. C. Abrahams, H. D. Flack, W. Gonschorek, Th. Hahn, K. Huml, R. E. Marsh, E. Prince, B. E. Robertson, J. S. Rollett and A. J. C. Wilson, *Acta Cryst.*, 1989, **A45**, 63.
[5] D. E. Sands, *J. Chem. Educ.*, 1977, **54**, 90.
[6] J. D. Dunitz, *X-Ray Analysis and the Structure of Organic Molecules*, Cornell University Press, Ithaca, 1979.
[7] R. Taylor and O. Kennard, *Acc. Chem. Res.*, 1984, **17**, 320.
[8] J. D. Dunitz, E. F. Maverick and K. N. Trueblood, *Angew. Chem, Int. Ed. Engl.*, 1988, **27**, 880.
[9] D. J. Watkin, *Acta Cryst.*, 2000, **B56**, 747.
[10] F. L. Hirshfeld, *Acta Cryst.*, 1976, **A32**, 239.

Exercises

13.1   Consider a set of supposedly equivalent bond lengths:

1.524	1.525	1.526	1.526	1.528	1.529	1.529
1.531	1.532	1.532	1.533	1.533	1.534	1.534
1.534	1.535	1.535	1.536	1.536	1.538	1.540

Calculate the mean bond length, the standard deviation of this set of values, and the standard deviation of the mean $\sigma(\bar{x})$.

13.2   For a function of independent variables x_i

$$\sigma^2(f) = \sum_{i=1}^{n} \left(\frac{\partial f}{\partial x_i}\right)^2 \sigma^2(x_i). \qquad (13.33)$$

Obtain $\sigma^2(f)$ for the functions
(a) $f = x_1 + x_2$
(b) $f = x_1 - x_2$
(c) $f = x_1 + x_2 + x_3$
(d) $f = x_1 x_2$.

13.3   Bond angles of a substituted cyclopropane ring, from the least-squares refinement, are 59.3(2), 59.6(2) and 61.0(2)°. What is the sum of the angles? What is its s.u.?

14

The interpretation of results

14.1 Introduction

So now we have a fully refined structure. The parameters obtained from the least-squares refinement are a set of coordinates and displacement parameters for each atom, and from these we are able to calculate geometrical parameters of interest: bond lengths, bond angles, torsion angles, least-squares planes with angles between them, intermolecular and other non-bonded distances. We can analyse the movement of the atoms and, perhaps, make some corrections to the apparent geometrical values we have calculated. To every derived result we can attach a standard uncertainty as a measure of its precision or reliability.

So far this has been largely straightforward mathematical calculation. But now we must begin to *interpret* the results, to detect patterns, common features, significant differences and variations, and to make deductions on the basis of the observed geometry. We shall need to compare features within the structure, and also compare them with other related structures.

In this chapter we concentrate on such questions of interpretation. How do we compare sets of results? How significant are differences between experimental values? How similar are two molecules? Just how approximate is approximate symmetry? How short should an interatomic distance be before we call it a bond? We shall not be able to give firm answers to all these questions!

We shall then consider what sort of effects various systematic errors may have on the precision and accuracy of our results, and examine some criteria by which we may judge the quality of other people's structural results – and our own!

14.2 Averages, comparisons and differences

We consider first the interpretation of a single structure, and then some additional points which arise in the comparison of different structures.

14.2.1 Comparison of geometrical parameters

A frequent question in interpreting a structure is whether a particular bond length or angle differs significantly from another bond length or angle, or from some standard value. Unless we make some assumptions about errors, we cannot deduce anything useful at all. The basic assumptions we make are (i) that each of our derived results is an *unbiased* estimate of its true value (i.e. we assume that we have an *accurate* result, either free from any significant effects of systematic errors, or corrected for them), and (ii) that the s.u.

is a true estimate of the precision of our result and a measure of the variation we would expect to find if we determined the structure many times; such a variation is expected to follow a normal distribution.

These assumptions, together with a knowledge of the properties of a normal distribution, allow us to work out the significance of the difference between two parameters. To compare two bond lengths, for example, we imagine that the true values are equal. If we were to measure the difference many times, these measurements would be subject to random error, and we would obtain a set of observed differences normally distributed about a mean of zero with some standard deviation. We have in fact only *one* measurement of the difference, together with an estimate of the standard deviation (by combining the s.u.s of the two bond lengths). The question now becomes: what is the probability that this difference, with a true value of zero, would be measured with such a value? The probability of finding a normally distributed variable more than $\pm\sigma$ from its mean value is 31.7%; the probability is 5% for a deviation outside $\pm1.96\sigma$, and 1% for a deviation outside $\pm2.58\sigma$. According to arbitrary limits commonly used in statistics, a probability of over 5% means the difference is not significant, and a probability of under 1% means that it is significant, with the 1–5% probability range being a grey area of 'possible significance'. Since it is widely believed that crystallographic s.u.s are over-optimistic (particularly as they assume only random errors), and since it is easiest to work with round figures, a difference greater than 3σ is often taken as a significant difference, while smaller values are probably not significant.

The above stated simply that the s.u. of the difference between two bond lengths (or other values) is obtained by combining the s.u.s of the two bond lengths. The correct expression for obtaining an s.u. of a function of two or more variables (see the previous chapter) is

$$\sigma^2(f) = \sum_{i,j=1}^{n} \frac{\partial f}{\partial x_i} \cdot \frac{\partial f}{\partial x_j} \cdot \text{cov}(x_i, x_j) \tag{14.1}$$

where $f(x)$ is a function of the variables $(x_i; i = 1, n)$. In comparing bond lengths or other geometrical parameters, we usually have only the variances (squares of s.u. values) and no covariances available. Ignoring covariance terms is equivalent to assuming zero correlation among the atomic positions, and gives the simpler expression

$$\sigma^2(f) = \sum_{i=1}^{n} \left(\frac{\partial f}{\partial x_i}\right)^2 \sigma^2(x_i). \tag{14.2}$$

In considering the difference between two bond lengths A and B, this becomes simply

$$\sigma^2(A - B) = \sigma^2(A) + \sigma^2(B) \tag{14.3}$$

and this is the formula generally used. Note, however, that the situation in which the bonds A and B are bonds to the same atom is one which definitely does *not* have zero covariances.

As an example, consider two bonds of length 1.540(3) and 1.570(4) Å: are they significantly different? The observed difference is 0.030 Å and the s.u. of this difference is

$(0.003^2 + 0.004^2)^{1/2} = 0.005$ Å. Therefore the difference is 6σ and this would certainly appear to be significant. On the other hand, by similar simple calculations, bonds of length 1.540(3) and 1.550(4) Å are not significantly different, and neither are bonds of length 1.540(8) and 1.570(9) Å. The importance of reducing random errors, as well as eliminating systematic errors as far as possible, is clear.

Comparison of a particular bond length with some 'standard value' is similar, except that the standard value is assumed to have a zero s.u., so the s.u. of the difference between the observed and standard value is the same as the s.u. of the observed value itself and the question becomes simply: does the value differ from the standard by more than three times its own s.u.? Thus, for example, a C–C bond length of 1.584(5) Å is significantly longer than a 'standard' value of 1.540 Å.

14.2.2   Averaging geometrical parameters

It is a common practice to average bond lengths which are supposed to be of the same kind in a structure (or in more than one structure). Two questions arise: (i) what is the best way to calculate the average, and (ii) is such an average meaningful?

An 'average' of values which do not all have equal precision may be calculated as a weighted mean

$$\bar{x}_w = \frac{\sum_{i=1}^n w_i x_i}{\sum_{i=1}^n w_i} \tag{14.4}$$

with $w_i = 1/\sigma^2(x_i)$, or it may be a simple unweighted mean

$$\bar{x}_u = \frac{\sum_{i=1}^n x_i}{n} \tag{14.5}$$

and the two means are, in general, not equal. In a detailed investigation of weighted and unweighted means [1], it has been shown that a weighted mean is appropriate if the variation in the values to be averaged is mainly due to experimental random errors, so that the observed values are normally distributed about their mean, which is the common 'true' error-free value for all of them. The unweighted mean should be used if the variation is mainly due to environmental effects such as crystal packing forces, so that the observed values really do represent genuine differences caused by physical factors perturbing them from their mean values. These may be described as 'hard' and 'soft' geometrical parameters respectively, and are typified by stiff bonds, on the one hand, and hydrogen bonding, on the other. If you are in doubt, the unweighted mean is usually acceptable in most circumstances.

For a set of values which really are equivalent and are normally distributed about their mean, this mean is more precisely determined than the individual values. The s.u. of the unweighted mean is given by

$$\sigma^2(\bar{x}_u) = \frac{\sum_{i=1}^n (x_i - \bar{x}_u)^2}{n(n-1)}. \tag{14.6}$$

For the weighted mean, the calculation is more difficult, because it should include both variance and covariance terms. Correlation should also be allowed for if bond lengths to be averaged involve common atoms, but this is overlooked more often than not.

For 'soft' variables a mean, whether weighted or not, is of rather dubious significance. Such is the case, for example, for bond angles, and even more so for torsion angles, and it may often be more useful and sensible to quote ranges of observed values rather than averages.

In many cases, the variation of a set of values is larger than would be expected from their individual s.u.s, and calculation of the s.u. of the mean value by the above method gives a conservative result, i.e. a generous and probably realistic estimate of the uncertainty. This is because individual s.u.s generally do not reflect the influence of systematic errors on the results of a crystal structure determination. Where the spread of values is comparable to individual s.u.s, the s.u. of the mean value can be obtained correctly by the methods shown in Section 14.2.1.

An average value will also be fairly meaningless for a set of parameters which are not really equivalent (i.e. do not belong to the same normal distribution). Even for bonds which appear to be chemically similar, statistical equivalence may not be found. A set of bond lengths (or other parameters) can be tested to see if they are probably drawn from a common distribution, by means of a statistical χ^2 test, in which the quantity

$$\chi^2 = \frac{\sum (x_i - \bar{x})^2}{\sigma^2} \tag{14.7}$$

is compared with standard χ^2 tables; this gives the probability of finding such a variation of values for a set of variables which are supposed to be all equal and are in fact normally distributed about their mean. The σ here is the overall s.u. of the variables, assumed in the test to be equal for each of them, so the test works well only for bond lengths of comparable precision (for which an average σ can be used).

Quite apart from such tests, a good helping of common sense is valuable! If two supposedly chemically equivalent metal–ligand bonds are found to have lengths of 1.925(2) and 1.985(2) Å, it seems pointless to quote an average bond length of 1.955 Å and argue about its s.u.

14.2.3   When is a set of atoms genuinely planar?

In the previous chapter we saw how a least-squares plane can be fitted to a set of atomic positions. The root-mean-square deviation of the atoms from the plane gives a measure of the planarity of the set. A χ^2 test can be applied to this quantity in a similar way to the one shown above, but it is in fact rare for the test to pronounce such sets to be truly planar.

A degree of interpretation is already effectively made when least-squares planes are calculated, because some sets of atoms are more obviously approximately planar than others, and we do not usually fit least-squares planes to random sets of atoms. The exact choice of which atoms to include in the calculation for an approximately planar set does make a lot of difference to the deviations found, as the least-squares method by its very nature will avoid a small number of large deviations. It may be more useful to fit a plane to a relatively small number of atoms which really are close to coplanar and examine the deviations of the other atoms from it than to fit a plane less well to a larger set of atoms; alternatively, two or more planes can be calculated for different subsets of the atoms and their dihedral angles listed. A good example is a non-planar four-membered ring, for which two exact three-atom planes with a hinge angle may be a more useful and revealing description than four deviations from a single plane, but it is impossible to generalize about the interpretation of results.

14.2.4   Comparing different structures

Comparisons of individual parameters of two or more different structures can be made in the ways described above. In addition, there are statistical methods available for comparing two structures wholesale.

(a) Normal probability plot analysis [2]
In the previous chapter we saw how two sets of data can be compared by a normal probability plot: the ordered values of the weighted deviations δ_i are plotted against the values expected for a normal distribution of zero mean and unit variance.

The same technique, with a small difference, may be used to compare two sets of atomic coordinates (or geometrical parameters derived from them). The difference is that the signs of the individual δ_i deviations are not important, and we use only the magnitudes $|\delta_i|$, where

$$|\delta_i| = \frac{\left|v_{1,i} - v_{2,i}\right|}{[\sigma^2(v_{1,i}) + \sigma^2(v_{2,i})]^{1/2}} \tag{14.8}$$

and the variable v_i stands for all three coordinates (x_i, y_i, z_i) for each atom, so for n atoms there are $3n$ deviations. These $|\delta_i|$, sorted into order from zero to maximum value, are plotted against the values expected for a standard half-normal distribution. A unit gradient and zero intercept indicate that the two sets of coordinates are essentially equivalent and their s.u.s are correctly estimated. A gradient greater than unity indicates that one or both sets of s.u.s are underestimated; systematic errors will produce non-linearity and, possibly, a non-zero intercept.

An example of the application of a half-normal probability plot to two independent determinations of the same structure is given in [3]. The plot is essentially linear, with zero intercept and a gradient of 1.39; the conclusion is drawn that the two sets of refined coordinates (derived from different sets of measured data) are not significantly different and that the s.u.s have been underestimated in one or both of the determinations (Fig. 14.1).

(b) Least-squares fit of two structures
Another method of comparing two structures is to attempt to superimpose the two molecules as closely as possible on each other. Essentially, this is achieved by referring the two structures to a common centre of gravity and then rotating one of them

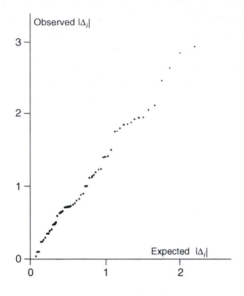

Fig. 14.1. A half-normal probability plot comparing two determinations of the same structure.

relative to the other about this fixed point to minimize the residual

$$\sum_{i=1}^{n} w_i d_i^2 \qquad (14.9)$$

where d_i is the distance between the corresponding ith atoms in the two structures; a total of n atom pairs are included in the fit. The weights may reflect the precisions of the individual atom positions (from s.u.s of coordinates) or may be chosen in some other way to attach greater importance to the fit of some atom pairs than others; often, however, unit weights are employed.

Various mathematical methods have been used to produce such a least-squares fit, and such a routine is available in several computer programs and packages.

The technique offers considerable flexibility and has many applications. Selected parts of two related molecules can be compared, for example when they have different substituents. Different parts of a molecule can be compared with each other to give an indication of approximate symmetry. An experimentally determined structure can be fitted to an ideal model. It is important to have a facility for inverting the handedness of one of the two structures if desired, in order to compare enantiomeric structures or to test for approximate reflection or inversion symmetry relating two parts of a molecule. An example is shown in Fig. 14.2.

(c) Limitations of inter-structural comparisons

Two detailed analyses of structural comparisons, with special reference to the Cambridge Structural Database, indicate some possible problems which may arise in comparing structures [4,5].

First, since structural parameters are usually reported to a certain number of decimal places, and s.u. values to only one or two figures, differences between pairs of these

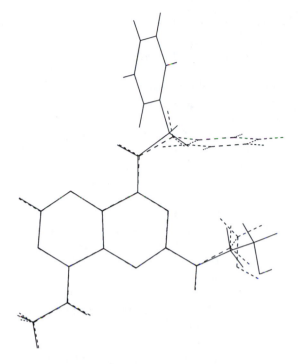

Fig. 14.2.   A least-squares superposition of two molecules, showing different conformations.

parameters will necessarily be rounded off to the same number of decimal places. If the differences are small, their distribution, therefore, will not be genuinely continuous, but will be, to some extent, a discrete distribution. This artefact of rounding may produce errors in the application of statistical tests [4].

Second, analysis of 100 pairs of independent determinations of the same structure [5] supports the widely held view that crystallographic s.u.s tend to be over-optimistically low: a factor of about 1.4 underestimation is indicated on average for the positional parameters (and consequently for the derived geometrical parameters) in these particular structures; the effect is greater for heavy atoms, and a much greater average factor (up to about 5) is found for cell parameters.

Thus, care must be exercised in making decisions about the significance of differences in geometrical values, and it is probably better to err on the side of caution, not claiming as genuine a difference whose significance (in a statistical sense) is marginal.

14.3   Interpretation of interatomic distances and bonds

In most organic molecules, bond lengths between most types of atoms fall within fairly narrow and well-defined ranges. Multiple bonds are shorter than single bonds, and bond lengths deviating considerably from 'standard' values are rare, except in very strained molecules (small rings, polycyclic molecules, those with extreme steric crowding, etc.).

In other branches of chemistry, bond lengths between atoms of the same pair of elements are much more variable, and formal partial bond order is a feature to be reckoned with. The variation of geometrical parameters within a structure, or between different structures, and deviations from 'standard' values can be assessed statistically as we have seen above. But how do we assess them *chemically*? At what distance do a pair of atoms have significant interactions, and what distance indicates a genuine bond?

These questions are traditionally answered by comparison with sums of some kind of atomic bonding radii. Thus a 'normal' bond length should be about equal to the sum of two covalent radii, and a significant non-bonded interaction is indicated by a distance shorter than the sum of relevant van der Waals radii. Suitable radii are available in various tabulations; the source most often referred to is probably [6].

There are, however, many problems associated with such decisions. First, it is not always clear which are the appropriate 'bonding radii' to use: depending on the particular material under investigation, they may be covalent, metallic or ionic radii, and these can differ enormously for some atoms.

Second, the radii are only approximate, and they are dependent on circumstances. It is well known, for example, that both ionic and covalent radii tend to increase with coordination number, so longer bonds are to be expected around an octahedral centre than a tetrahedral one, for example. Radii also vary with oxidation state [e.g. Fe(III) is smaller than Fe(II)] and with electron configuration [e.g. high-spin Ni(II) is considerably larger than low-spin Ni(II)]. It has been demonstrated that van der Waals radii can be markedly anisotropic [7], so that a distance that represents a significant interaction in one direction may be nothing remarkable at all in another direction in the same crystal structure.

Third, and not surprisingly in view of some of the above comments, these radii are by no means always certain, and considerable variation is found in published tabulated values. A good example is the van der Waals radius for mercury, for which a wide variety of values has been proposed. It is unwise to mix values of van der Waals radii from different compilations, as these may not be consistent with each other.

It is particularly among heavier elements that considerable variation of interatomic distances is found, with a wide range of strength of secondary interactions as well as direct primary bonding.

Attempts to bring some quantitative order to this difficult area of interpretation include formal definitions of coordination number, and various measures of bond valence and other bond strength criteria. Bond length–bond order correlations are, however, notoriously unpredictable and wayward.

14.4 The effects of errors on structural results

We saw in the previous chapter that random errors in the experimental data affect the precision, but not the accuracy, of the results. Systematic errors may or may not affect the precision, but they almost always affect the accuracy. Thus, reducing random errors will give smaller s.u.s in the final results, and reducing systematic errors will give final results which are closer to their 'true' values.

The final results are affected not only by the data themselves but also by how they are used. The choice of least-squares refinement parameters, the method of refinement, and the weighting scheme all have an effect on the geometry we are trying to determine.

In this section, we consider the effects of some errors, particularly systematic errors, on the results of a crystal structure determination. To some extent this overlaps with previous chapters on data collection and refinement, and acts as a summary.

14.4.1   Systematic errors in the data

(a) Absorption
Absorption reduces the observed intensities of diffraction, but by different factors for different reflections. The effect is greatest at low Bragg angle, so that there is a systematic error even for a spherical crystal. Uncorrected significant absorption causes atomic displacement parameters to be too low, in an attempt to compensate for the effect. Anisotropic absorption (for a non-spherical crystal) affects the apparent atomic vibration differently in different directions, so that distorted 'thermal ellipsoids' are produced for the atoms in a needle or plate crystal. The atomic coordinates are generally not significantly affected, but the s.u.s are increased because the observed and calculated data do not agree so well; the atomic displacement parameters cannot completely mop up the absorption errors.

(b) Extinction
This also attenuates the observed intensities, and it is most severe for low-angle, strong reflections. Like absorption, it reduces overall precision and systematically affects atomic displacement parameters, while having much less effect on atomic coordinate values.

(c) Thermal diffuse scattering
Thermal diffuse scattering (TDS), produced as a result of cooperative lattice vibrations, has the effect of increasing observed intensities. The effect, however, increases with $\sin^2\theta$, so the net effect, if no correction is made, is once again to reduce atomic displacement parameters from their true values. TDS effects have received little attention in routine crystal structure determination, and they are *generally* believed to be small. Data collection at reduced temperature is an advantage here, as well as in other ways.

(d) A poorly aligned diffractometer
Several errors can be introduced into the diffraction data by this fault, including an improper measurement of intensities if the reflections are not completely received by the counter aperture. The most common error, however, is probably in unit cell parameters. If a badly aligned instrument involves systematic errors in the zero-points of the circles (especially 2θ), there will be a corresponding error in refined cell parameters, which may well be much greater than their supposed s.u.s. This, in turn, leads to systematic errors in molecular geometry, and no indication of these can be seen in the commonly quoted measures of the 'quality' of the structure determination (structure factor residuals, goodness of fit, etc.), which refer only to diffraction intensities and not to diffraction

geometry. Such errors as this, therefore, are the most invidious, because it is difficult to detect them.

(e) Anomalous dispersion

This may be considered as a systematic effect (and, hence, a potential source of systematic error) in the data, or as a possible fault in the structural model if properly corrected atomic scattering factors are not used. Neglect of the correction in a non-centrosymmetric structure with a polar axis or, even worse, a 'correction' with the wrong sign, results in a systematic shift, along the polar axis, of all atoms displaying significant anomalous scattering effects, because this shift, relative to the rest of the atoms, mimics the phase shift produced by the anomalous scattering [8]. In a centrosymmetric or a non-polar non-centrosymmetric structure, atomic positions are not affected but atomic displacement parameters are.

Of course, anomalous dispersion effects can be used for determining the correct 'handedness' of a non-centrosymmetric structure (see Chapters 12 and 18), and also for determining phases of reflections in structure solution, but these are really separate subjects and not concerned directly with errors and results.

14.4.2 Data thresholds and weighting

It is fairly common for the weakest reflections not to be used in least-squares refinement, though there is considerable controversy over this. The inclusion or omission of weak reflections usually makes no significant difference to the derived parameter values. Weak reflections tend to increase the residuals R and (to a lesser extent if they are correctly weighted) wR, but this is compensated somewhat by the larger excess of data over parameters $(N - P)$, so it is not clear how final s.u.s will be affected. In fact, it has been demonstrated that they, too, are scarcely altered by the inclusion or omission of weak reflections or by the decision of just where to set the threshold [9]. Thus, the difference in the results is to a large degree just a cosmetic one. On the other hand, the weak reflections can play a crucial role in deciding between centrosymmetric and non-centrosymmetric space groups in ambiguous cases, because their omission tends to bias statistical tests towards a decision against centrosymmetry.

We have already seen, in the previous chapter, ways of checking for a reasonable weighting scheme through various analyses of the agreement between the observed and calculated data. An inappropriate weighting scheme will affect final results in rather unpredictable ways, and it will generally increase s.u. values.

14.4.3 Errors and limitations of the model

The parameters refined by least squares are an attempt to describe the structure we are trying to determine. They represent an approximation to the actual X-ray scattering power of the structure. No such model can be a perfect representation, and there are various limitations on the simple models we use, and various errors that may be made in choosing the elements of the model.

(a) Atomic scattering factors

The tabulated scattering factors commonly used are reasonably accurate representations of the scattering power of individual, isolated atoms at rest. They have spherical symmetry, and probably their greatest limitation is the lack of allowance for distortion of this spherical electron density when atoms are placed together and bonded to each other. The greatest effects of this approximation are seen in the low-angle data, and an analysis of variance of the observed and calculated data after refinement commonly shows the worst agreements for such reflections. In careful work, some of the largest peaks in a final difference electron density synthesis are found between atoms (bonding electron density) and in regions where lone-pairs of non-bonding electrons are expected to lie (Fig. 14.3). This is, of course, one reason why bonds to hydrogen atoms are found to be systematically shortened in X-ray diffractions studies.

An incorrect assignment of atom types, so that a wrong scattering factor is used for an atom, scarcely affects atomic coordinates in most cases, although there are circumstances under which these may be subject to a systematic error. In an attempt to compensate for the wrong scattering factor, refinement will adjust the atomic displacement parameters, often to a very considerable degree; an incorrectly assigned atom may be recognized in many cases by its anomalous displacement parameter, especially at the early isotropic stage of refinement when there is a single parameter for each atom.

(b) Constraints and restraints

Properly used, these are valuable tools in refinement, allowing us to deal with problems of parameters which are not well determined from the diffraction data alone. We must,

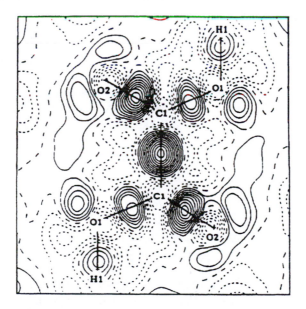

Fig. 14.3.   Valence electron density, seen as deviations from assumed spherical atomic scattering factors.

however, be quite sure of the validity of any constraints or restraints we apply. Any which strongly oppose the course of unconstrained/unrestrained refinement, rather than gently guiding it, prevent convergence to the data-determined minimum and so force a different result. This will, in particular, significantly affect the geometry around the regions in the structure where the constraints are being applied.

A common example is the use of a constrained C–H bond length of 1.08 Å, chosen because it is the 'true' value determined spectroscopically for simple hydrocarbons. Since C–H bonds are systematically shortened in X-ray work, the effect of the constraint will be to push both atoms further apart than the diffraction data alone would indicate. Although the hydrogen atom position will be most affected, there will be a small, but possibly significant, effect on the carbon atom. Misplacing the atoms in this way will also affect their displacement parameters.

Other cases of inappropriate constraints include the imposition of too high a symmetry on a group of atoms which is genuinely perturbed to a less regular shape by bonding or packing interactions. Phenyl groups are systematically distorted from regular hexagonal symmetry, and the use of such a simple model, frequently used in refinement, may not be appropriate.

(c) Incorrect symmetry

Space group determination is based on several experimental measurements and deductions: (i) the metric symmetry of the reciprocal and direct lattices; (ii) the Laue symmetry of the observed diffraction pattern; (iii) systematically absent reflections; (iv) statistical tests for the presence or absence of symmetry elements, especially an inversion centre; (v) ultimately, a 'successful' refinement. Reports appear relatively frequently in the literature of space groups which are reputed to have been incorrectly assigned by previous workers. In many cases, the problem is not a serious one, in that two molecules, equivalent by unnoticed symmetry, are refined as independent, and their geometries are not significantly different: the results are reliable, but contain unnecessary redundancies. Where the missing symmetry includes an inversion centre, however, there is a real problem, in that refinement is unstable (strictly speaking, the matrix is singular), but this may be masked by the particular refinement technique used. The geometrical results in this case are often unreliable: parameters which should be equal by symmetry may be found to differ by a large amount, and the molecular geometry often displays considerable distortions. These may be hidden by imposed constraints or restraints.

(d) High thermal motion and static disorder

It is not always easy to distinguish these two situations, except by carrying out the data collection at a reduced temperature (which reduces dynamic disorder but not usually static disorder unless there is an order–disorder phase transition at an intermediate temperature). High thermal motion increases the foreshortening of interatomic distances generally observed in X-ray diffraction, so there is a considerable systematic error in bond lengths, which was discussed in Chapter 13. The usual six-parameter (ellipsoidal) model of thermal motion becomes increasingly inadequate as the motion increases in amplitude, so the displacement parameters are of dubious value and their precision is generally poor.

The presence of disorder in a structure, unless it is very simple and can be well modelled, reduces to some extent the overall precision of the whole structure, not just of the particular atoms affected. For this reason, certain atomic groupings notorious for disorder are best avoided if possible: these include ClO_4^-, BF_4^- and PF_6^- anions.

High thermal motion and/or disorder can make the geometrical interpretation of a structure difficult, and may lead to incorrect deductions about the molecular geometry and conformation. A classic case is that of ferrocene, $(C_5H_5)_2Fe$, which appears to be staggered because of unresolved disorder at room temperature, but which (contrary to statements in some standard inorganic chemistry textbooks) is in fact eclipsed [10].

(e) Wrong structures

Such errors as those just mentioned, with an incorrect molecular geometry, are bad enough, but it is possible, though very uncommon and unlikely, to find a completely incorrect structure, in the sense of identifying the wrong chemical compound. A case of mistaken identity caused by 30-fold disorder involves the supposed structure of dodecahedrane [11]. Wrongly assigned atom types were suspected in the structure of '$[ClF_6][CuF_4]$', which in reality is probably $[Cu(H_2O)_4][SiF_6]$ [12]; the original workers were misled by the similarity in scattering powers of Si and Cl, and of O and F, and perhaps by some wishful thinking.

14.5   Assessment of a structure determination

The above discussion should encourage us to take a critical view of crystal structure determination in general [13], and to seek to evaluate carefully any particular reported structure. Several research journals issue detailed instructions for authors of crystal structure reports, and some provide separate checklists for referees. These can provide a useful framework for assessing a structure, whether it be one reported in the literature, or one of your own.

Below is a summary of some useful points for checking the quality of a structure determination. It is derived from a number of sources, including standard tests applied by *Acta Crystallographica Sections C* and *E*, and a list distributed by David Watkin at a British Crystallographic Association Intensive School.

1. Check for consistency of the crystal data. If you have a suitable computer program, you can input the cell parameters and chemical formula and check the volume, Z (number of chemical formula units in the unit cell), density, absorption coefficient μ, etc. For a quick check by hand calculation:

(a) count the non-hydrogen atoms (N) in the molecule or formula unit;
(b) check that $abc \sin \delta \approx V$, where δ is the most removed of α, β, γ from 90°;
(c) calculate the average volume per non-H atom ($= V/NZ$), which is usually about 18 $Å^3$ for organic and many other compounds.

2. Assess the description of the data collection.

(a) Check that $|h|_{max}/a \approx |k|_{max}/b \approx |l|_{max}/c$, and that at least the correct minimum fraction of reciprocal space has been covered.
(b) $2\theta_{max}$ should be at least 45° (better, 50°) for Mo radiation, 110° (better, 130°) for Cu radiation.
(c) Look at the number of unique data, the number of 'observed' data, and the threshhold [which may be expressed in terms of $\sigma(I)$ or $\sigma(F)$: $I \geq 2\sigma(I)$ corresponds to $F \geq 4\sigma(F)$]; check for a low value of R_{int} if equivalent reflections are merged.
(d) Calculate and compare μt_{min} and μt_{max} (dimensionless!) for the minimum and maximum crystal dimensions. If $\mu t_{max} < 2$, absorption is probably no problem. If $\mu t_{max} > 5$ or $(\mu t_{max} - \mu t_{min}) > 2$, an absorption correction is necessary (or there will be significant effects on the U^{ij} values). More detailed tests are described in the *Notes for Authors* of Acta Crystallographica Section C.

3. Assess the refinement and results.

(a) The number of *observed* data should be greater than the number of refined parameters by a factor of at least 5 (and preferably 10). Anisotropic refinement gives 9 parameters per atom, isotropic gives 4. *Constrained* H atoms do not count unless U is refined for them).
(b) Examine carefully the description of any constraints/restraints, the treatment of H atoms, and any disorder.
(c) Look for strange U or U^{ij} values; high values may indicate disorder, low values may indicate uncorrected absorption (unless low temperature was used); either of these could be due to misassigned atom types.
(d) Check for convergence (shift/s.u. values preferably <0.01).
(e) Examine the fit of observed and calculated data, if possible, not only by the value of R.
(f) Difference electron density outside about ± 1 e Å^{-3} may be due to missing, misplaced or misassigned atoms, systematic errors such as absorption (especially if large peaks appear close to heavy atoms), or unmodelled disorder.
(g) Check for 'absolute structure' determination if the space group does not have a centre of symmetry.
(h) Assess the s.u.s (i) of the refined parameters; (ii) of the molecular geometry parameters. Watch out for low s.u.s ignoring cell parameter uncertainties. Check for s.u.s on values which should be constrained, symmetry-equivalent, etc.
(i) Check for strange results: unusual geometry, impossibly short intermolecular contacts, etc.

Two accounts of 'problem structures' are also given in [14]. The use of a comprehensive checking and analysis program such as PLATON [15] is strongly recommended.

References

[1] R. Taylor and O. Kennard, *Acta Cryst.*, 1983, **B39**, 517.
[2] S. C. Abrahams and E. T. Keve, *Acta Cryst.*, 1971, **A27**, 157.

[3] W. Clegg, N. Mohan, A. Müller, A. Neumann, W. Rittner and G. M. Sheldrick, *Inorg. Chem.*, 1980, **19**, 2066.

[4] R. Taylor and O. Kennard, *Acta Cryst.*, 1985, **A41**, 122.

[5] R. Taylor and O. Kennard, *Acta Cryst.*, 1986, **B42**, 112.

[6] L. Pauling, *The Nature of the Chemical Bond*. Third edition, Cornell University Press, Ithaca, 1960, pp. 221–264.

[7] S. C. Nyburg and C. H. Faerman, *Acta Cryst.*, 1985, **B41**, 274.

[8] D. W. J. Cruickshank and W. S. McDonald, *Acta Cryst.*, 1967, **23**, 9.

[9] R. E. Stenkamp and L. H. Jensen, *Acta Cryst.*, 1975, **B31**, 1507.

[10] E. A. V. Ebsworth, D. W. H. Rankin and S. Cradock, *Structural Methods in Inorganic Chemistry*, 2nd edn, Blackwell, Oxford, pp. 414–417.

[11] O. Ermer, *Angew. Chem., Int. Ed. Engl.*, 1983, **22**, 251.

[12] H. G. von Schnering and Dong Vu, *Angew. Chem., Int. Ed. Engl.*, 1983, **22**, 408.

[13] P. G. Jones, *Chem. Soc. Rev.*, 1984, **13**, 157.

[14] J. A. Ibers, *Problem crystal structures* and J. Donohue, *Incorrect crystal structures: can they be avoided?* In *Critical Evaluation of Chemical and Physical Structural Information*, ed. D. R. Lide Jr. and M. A. Paul, Nat. Acad. Sci., Washington DC, 1974.

[15] PLATON, a program for the automated analysis of molecular geometry. A. L. Spek, University of Utrecht, The Netherlands.

Exercises

14.1 Are the following parameters significantly different?
(a) bond angles of $110.3(2)°$ and $110.5(2)°$;
(b) bond angles of $93.2(5)°$ and $95.7(4)°$;
(c) a bond angle of $177(1)°$ and linearity.

14.2 Assume all the bond lengths in Exercise 13.1 of Chapter 13 have an s.u. of $0.003\,\text{Å}$. Do any of the bond lengths differ significantly from the mean? To answer this, calculate $\chi^2 = \Sigma(x - \bar{x}^2)/\sigma^2$. For 20 degrees of freedom, tabulated χ^2 values are 31.41 for a probability (significance) level of 5%, and 37.57 for a level of 1%; from these figures, are the bond lengths all equivalent?

14.3 Which of these symmetry elements make a four-membered MLML ring strictly planar? In each case, how many bond lengths are independent?
(a) a centre of symmetry;
(b) a twofold axis normal to the mean plane of the ring;
(c) a twofold axis through the two M atoms;
(d) a mirror plane through the M atoms but not through the L atoms;
(e) a mirror plane through all four atoms.

14.4 A six-coordinate atom lies on an inversion centre. How many independent bond lengths and angles are there around this atom?

14.5 Repeat Exercise 14.4 for the case where there are no symmetry constraints on any of the atoms.

15

The presentation of results

15.1 Introduction

The final stage of a crystal structure analysis is its presentation. This can occur within your own research group, as a conference poster or oral contribution, on the Internet, or as a refereed article in a journal. In each case the requirements are different and you must tailor the presentation to the medium used. As with all communication skills, the presentation of crystallographic results improves with practice. The reporting of structural results in crystallographic or chemical journals is usually guided by the relevant *Notes*, *Instructions* or *Guidance for Authors* published in these journals. A small number of journals only accept electronic submissions, that is by e-mail, through web interfaces, or on diskette, in a specific computer-readable format. Since April 1996, *Section C* of *Acta Crystallographica* only accepts submissions as Crystallographic Information Files (CIFs), and the same is true of the new *Section E*; *Zeitschrift für Kristallographie*'s New Crystal Structures section accepts CIF or CASTOR formats.

Even among crystallographic journals there has been a marked trend away from publishing primary data (coordinates and displacement parameters). Greater selectivity in the choice of molecular geometry parameters to be published is also being encouraged. Some journals published extensive information but did not always convey the structural information effectively, and now it is even more important to include effective graphical representations of your structures. With the spread of graphical abstracts in the contents pages of journals, a picture which is clear and attractive can be effective in attracting the attention of a reader browsing a hard-copy journal or an index on the WWW.

This section will deal first with molecular graphics, then with the production of tables, and finally with the different methods of delivering your results. Archiving will also be mentioned briefly. The use of the CIF will be referred to here as necessary, but will be covered in more detail in Chapter 16.

15.2 Graphics

Although most obviously associated with the production of high-quality views of the final structure, molecular graphics are also used as an aid in initial structure determination, in the interpretation of difference electron density maps, and to investigate disorder and other situations requiring modelling. Here we are only concerned with the first of these, and the main consideration is the quality of the resulting illustration in terms of its clarity, effectiveness and information content. Early graphics programs (e.g. [1,2]) were not interactive and the program had to be re-run every time a new view was required. Fortunately, modern programs are interactive (e.g. [3–5]), allowing continuous or stepped rotation of the molecule.

15.2.1 Graphics programs

The range of graphics programs is vast and it is not practical to offer a comprehensive survey here (one source of information on potentially useful programs is the website of the Collaborative Computational Project CCP14 at www.ccp14.ac.uk). The majority are free (at least to academic users) or cost very little. A major factor in your choice of program is its range of features and how these match your needs. Are atomic displacement ellipsoids required? Are polyhedral representations important? Do you want to display a ball-and-stick drawing of a molecule within its van der Waals envelope? How easy is it to deal with symmetry operators if you have molecules in special positions or if you want packing diagrams? A further point concerns the ability of the program to read and write data in certain formats. For example, if you regularly need to represent the data in files from the Cambridge Structural Database it is desirable that your program can do this without manual editing of the input file. More generally, a program's ability to generate plot files in standard formats such as HPGL or Postscript makes it possible to incorporate these into documents or transmit them by e-mail or FTP, or to a networked printer or plotter. If a program cannot export its output in some useful format, you are dependent on the quality of its output to a local printer or plotter. Note that programs are available which read and convert different data formats, e.g. BABEL, and different graphics formats, e.g. GHOSTVIEW.

There are also commercially available programs, usually supplied by diffractometer manufacturers as part of a complete package for structure analysis, but in some cases these can be purchased separately from the instrumentation. Such packages have the advantage of integration: for example, the solution and refinement programs communicate directly with the graphics module, and the problems that can arise due to incompatible data formats are avoided. Integration is also available in freely available software (e.g. WinGX [6], which provides a graphical interface linking various public-domain programs).

The increasing power of desktop computers and the availability of cheap laser printers with 600 dpi or higher resolution has meant that publication-quality illustrations can be produced with what is now very standard and affordable hardware. If you are considering the purchase of a system for structure analysis, the advice is the same as for any computer-related purchase. First choose the software you need, then select hardware you know will run it. Buy the fastest processor (for structure solution and refinement), the best screen (for viewing the structure), and the best printer (for hardcopy output) you can afford.

15.2.2 Underlying concepts

The positions of the atoms in a structure are derived from their fractional coordinates (x, y, z) on the crystal unit cell axes. Any graphics program needs to read these, along with the cell parameters required to convert to the orthogonal coordinate system in which the necessary calculations will be performed. The actual orthogonal axis set (x_o, y_o, z_o) is arbitrary and is not important. If the program accepts and can use crystallographic symmetry operators, it is necessary to read in only those atoms comprising the crystallographic asymmetric unit: symmetry-equivalent parts of the structure, whether for a molecule straddling a special position or for a packing diagram, can then be generated by the program. If the program cannot handle symmetry, then all the atoms required for the drawing must be generated before being input. (One reason for doing this could be to exploit a particular drawing style not available in the graphics routine you normally use.) For the drawing itself the program uses a separate coordinate system (x_p, y_p, z_p) in which the axes are defined relative to the drawing medium (usually a screen), and coordinates are generated only in order to produce the plot. The axes of this coordinate system are variously defined in different programs and this represents a minor source of possible confusion if you use a number of these.

The rotation (or view) matrix transforms the initial arbitrary view defined by the orthogonal coordinates (x_o, y_o, z_o) into the plotting coordinates (x_p, y_p, z_p) corresponding to the viewing direction required. Much of the ease of use of a program is associated with the flexibility and simplicity with which this can be done. The following options may be available:

(a) direct input of the nine elements of a rotation matrix (of limited interest);
(b) a view along cell axes or other crystallographic directions;
(c) a view with respect to molecular features, for example
    (i) along the direction perpendicular to the mean plane through selected (or all) atoms;
    (ii) along the vector between two atoms (which need not be bonded).

A good general approach is to start by looking along the direction perpendicular to the mean plane through all the non-H atoms, then make small rotations to refine this view. It is always a good idea to explore a range of views, in case a less obvious one proves to be the best. Most programs will allow you to do this either by continuous rotation or in small incremental steps, and unless you have a very large structure or a slow computer, the default rotation speed should be acceptable. In fact, with faster processors, you may need to slow the rotation rate for smaller molecules to prevent their spinning too rapidly.

Most drawings are composed of atoms and the bonds linking them. The connectivity information, which tells the program which atoms to draw bonds between, can be input explicitly along with the coordinates, but it is more common for the program to calculate the connectivity array using values for covalent or other radii appropriate to each atom. For example, the program may consider that two atoms are bonded if their separation is less than the sum of their covalent radii (plus a 'fudge factor' to ensure that slightly longer bonds are not missed): if the stored default covalent radius for carbon is $0.70\,\text{Å}$ and the default 'fudge factor' is $0.40\,\text{Å}$ then any pair of carbon atoms will be deemed to be bonded if they are within $0.70 + 0.70 + 0.40 = 1.80\,\text{Å}$ of each other. This approach is generally valid for organic compounds, where covalent radii are well defined, and many users may be unaware of the default values, simply because they never need to change them. With organometallic and inorganic compounds more care must be taken to ensure that all relevant interatomic distances are considered, but no unwanted ones. It may be necessary to edit the connectivity list in order to add or remove specific entries. There will be limits on the numbers of atoms and bonds which can be handled within any particular program; these limits may be set within the program, perhaps at compilation, or they may be determined by the memory available.

15.2.3 Drawing styles

A wide range of representations is possible, and it is important to choose appropriately. The simplest is a stick drawing (Fig. 15.1), where bonds are represented by straight lines; atoms are implied by bond intersections or termini. Some programs use this representation for rapid preliminary assessment of the best viewing direction, as it is the least demanding in terms of computing power (and is therefore fastest). It may also be the best way to display some complex molecules such as steroids, where drawing atoms as spheres or ellipsoids would obscure the features behind them.

A more usual style (ball-and-spoke, Fig. 15.2) involves displaying atoms as spheres (circles in projection), with the bonds shown as cylindrical rods. The user can select the

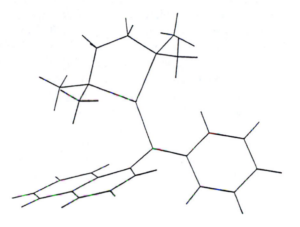

Fig. 15.1. A stick diagram.

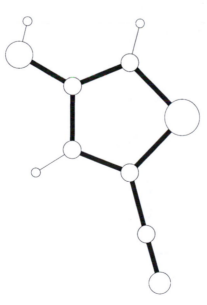

Fig. 15.2. A ball-and-spoke diagram.

radii of the spheres and the width of the bonds, alter the bond style, and add shading or other effects to the atoms; Fig. 15.3 shows the styles available in SHELXTL/PC version 5.03 [3]. Such features can be used to emphasize features of importance, such as the coordination sphere in a metal complex (Fig. 15.4). Different bond types can be used to indicate π-bonded ligands in metal complexes, or interactions such as intramolecular hydrogen bonds.

A highly informative type of plot, colloquially referred to as an 'ORTEP' after its historically best-known implementation [1,2], depicts atomic displacements as displacement ellipsoids. If the program offers a range of ellipsoid styles (e.g. those labelled -4 to -1 in Fig. 15.3), it will be possible to represent atomic motion and (to a limited extent) differentiate atom types in the same drawing (Fig. 15.5). The ellipsoids can be scaled in size to represent the percentage probability of finding within them the electrons around the atom as it vibrates; this probability level must always be quoted in the figure caption, and a value of 50% is typical, although values such as 20% or 70% are sometimes used to obtain reasonable views of structures with particularly high and low U^{ij} values, respectively. This type of plot is exceptionally helpful in highlighting possible problems such as disorder which may not be obvious from the numerical U^{ij} values, even when these are available. In fact, some crystallographers distrust ball-and-spoke plots, because disorder and other potential problems such as incorrect atom assignments can remain hidden.

In reality, molecules do not consist of balls and spokes, and these give a poor idea of the external shape and steric requirements of a molecule. A more realistic representation is provided by a plot such as Fig. 15.6, where the atoms are shown as spheres having

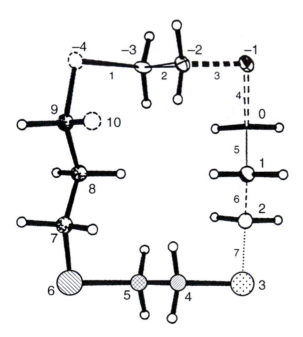

Fig. 15.3. Styles of atoms and bonds available in SHELXTL/PC [3].

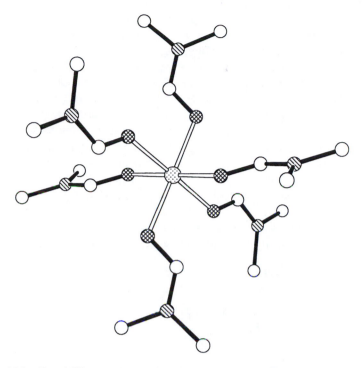

Fig. 15.4.   Use of different styles to differentiate atoms and emphasize particular features.

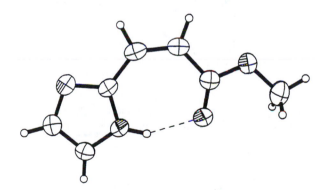

Fig. 15.5.   Displacement ellipsoid plot with a 50% probability level for the non-hydrogen atoms.

van der Waals radii rather than much smaller, arbitrary radii. These are referred to as 'van der Waals' or 'space-filling' plots and can be used to investigate questions such as whether a central metal atom is fully enclosed by its ligand array or is exposed and therefore more likely to undergo reaction. The program SCHAKAL [7] lets you display a composite picture of a ball-and-stick drawing of your molecule within its van der Waals envelope.

In addition to views of individual molecules, it may be desirable to show a number of molecules to demonstrate the relationships between them; interactions may involve hydrogen bonding or other secondary bonding, $\pi-\pi$ stacking of aromatic systems, or weak interactions such as those implicated in liquid crystalline behaviour. Such illustrations are subsumed under the name 'packing diagrams', an example of which is shown in Fig. 15.7. The molecules to be included can be selected automatically by the program (using criteria such as distance from a reference point or a certain number of unit cells), but the default selection may not give the best diagram. It is important to select and design packing diagrams in order to bring out the points you wish to illustrate without introducing unnecessary clutter. For this reason, the alternative approach of explicitly generating the required symmetry-equivalent molecules by the use of symmetry operations and cell translations has much to recommend it, even although it demands a higher level of understanding of crystal symmetry and expertise in using the more advanced features of graphics programs. Packing diagrams are frequently of poor quality and low information content, and suggest that to the maxim 'one picture is worth a thousand

Fig. 15.6. A space-filling plot.

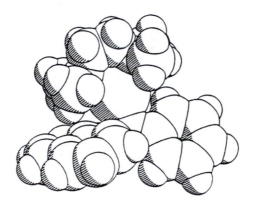

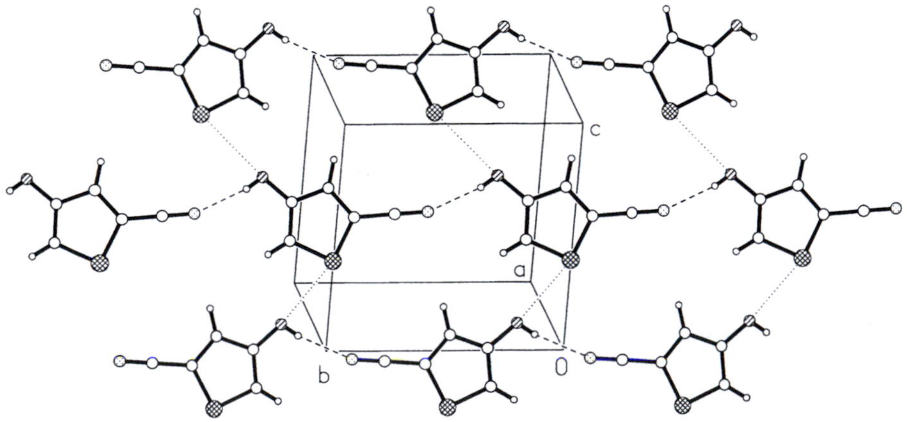

Fig. 15.7. A packing diagram.

words' should be added 'but only if it is a good one'. If the point of your packing diagram is merely to show that your molecules form typical linear chains it may be more efficient to convey this in words. With some journals adopting a policy that only one illustration of a structure is normally published, the ability to produce plots with high information content is very useful. For example, it may be possible to show both a single molecule and the salient features of its environment in a single illustration (e.g. Fig. 15.8) rather than as two separate ones. The important interactions between molecules should be made as clear as possible, typically by the use of different bond types; for example, in a structure containing two types of hydrogen bond these could be differentiated using dashed and dotted lines, while normal (intramolecular) bonds are shown by solid lines. If your figure is to be a representation of the full crystal structure, you will probably want to include the outline of the unit cell, with the axes labelled. The labelling of symmetry-related atoms is a slightly difficult area, as the method recommended by some journals (superscripted lower-case Roman numerals) is not easily available in many graphics programs. Fortunately, it is usually possible to work round this with a little ingenuity.

The captions to packing diagrams are probably one of the least exploited ways to convey structural information. They are often limited to 'Fig. 2: A view of the crystal packing', when they could contain concise information about the view direction, the most important contacts and distances, and the resulting arrangement of molecules (see Fig. 15.7).

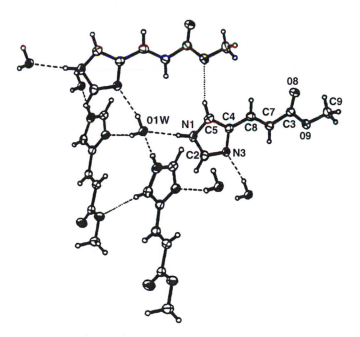

Fig. 15.8.   A composite diagram showing both a single molecule and its environment.

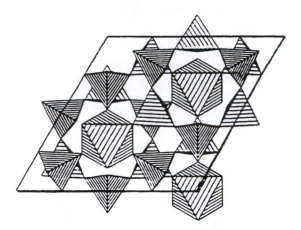

Fig. 15.9. A polyhedral
representation.

If you work with inorganic compounds, molecular representations may be less appropriate than polyhedral plots in which groups of atoms form polyhedral shapes (e.g. six
oxygens around a metal centre might be linked to generate an octahedron), which are
shown as opaque solid shapes. Neighbouring polyhedra are linked through their vertices,
edges or faces to build up the structure. As with packing diagrams of molecules, the selection of suitable symmetry-equivalents is important to the effectiveness of the illustration.
Figure 15.9 is a view of an inorganic structure produced by the STRUPLO program [8].

15.3   Creating three-dimensional illusions

When even a simple three-dimensional structure is represented in two dimensions there
is loss of information and for more complex cases this can be quite misleading. Various
tricks have been developed to give an illusion of depth on the screen or on paper, including
depth cueing, the use of perspective, bond tapering, hidden line removal and the use of
shading and highlights. Some techniques (e.g. depth cueing) work better on the screen
than on paper because of the different background colours used. You may find that your
graphics program has the required parameters already optimized, but conversely it may
be rewarding to adjust these so as to produce the best effects for your example. However,
beware of overdoing effects such as perspective or bond tapering to such an extent that
the result looks ridiculous.

The traditional way to restore some depth to a flat molecular plot is by inducing
stereopsis. The 'stereo pair' (Fig. 15.10) consists of two drawings, one for each eye, with
a suitable separation and slightly different rotations. When viewed, the two drawings
should merge to give a complete three-dimensional effect. Such pairs are not universally
effective, as a substantial proportion of the population literally cannot see the point of
them (these people cannot get the two images to merge) and their effectiveness should
be compared with that of a 'mono' plot occupying a similar area.

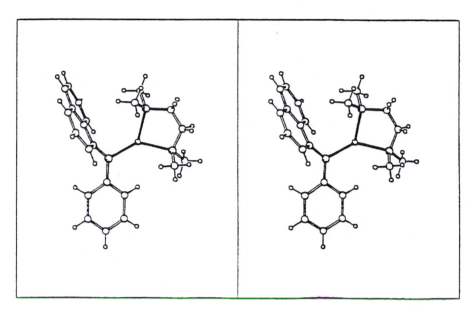

Fig. 15.10.   A plot consisting of a stereo pair.

15.4   The use of colour

In the past few years colour plots of crystal structures have become much more accessible, due to the wider availability of suitable software, the falling price of high-quality colour printers and the greater readiness with which many journals will publish colour plots, often at no charge but only where referees are convinced that they enhance the presentation significantly. The use of colour is most effective where it illuminates features which could not otherwise be easily identified: for example, in a polymeric structure two different metal atoms may have similar-looking high coordination and it might be impossible to find room for adequately sized labels to differentiate them. Colour codes can be defined in the figure caption or by means of a key panel. Other applications include colour coding of different bonds or atoms; highlighting of important features; colour coding of different molecules or structural motifs such as planes in packing diagrams; and sometimes just because colour looks wonderful on a poster. As with many good things, there are advantages in moderation: overuse of colour can be distracting and, if all or most of the atoms are coloured, many of the benefits of its use may be lost. Note that there are certain loose conventions about atom colours which you can use to convey more information: orange for B, black for C, light blue for N, red for O, light green for F, brown for Si, yellow for S, a darker green for Cl, and blue, green or red for metals. You are under no obligation to use these colours, but using other colour schemes will confuse at least some of those looking at the plot. Colour is probably least effective if only thin atom outlines are coloured and the colours are weak: it is better to fill the atom with a strong, vibrant colour. The use of colour opens up the possibility of using variations in intensity to convey depth information (depth cueing).

15.5   Textual information in drawings

Although the main constituents of a molecular plot will be atoms and bonds, it is normal (but not always essential) to include text, most commonly atom labels such as C1, N(2) or O3 (e.g. Fig. 15.8), or atom types (C/N/O). Unless colour or shading has been used to identify atom types, a drawing consisting only of chemically indistinguishable atoms and bonds is of limited use. Make sure that any labels are of a sufficient size that they will be legible at their final reduced size (but not so huge that they overwhelm the structure) and placed so that they will not overlap or merge with any atoms or bonds. Obviously, make sure the labels refer to the correct atoms and do not leave any ambiguity. If the graphics program inserts them automatically, do check their placement. One decision is whether to have parentheses in your atom labels or not [e.g. C24 or C(24)]: the latter require more space but in some circumstances can help to avoid confusion [e.g. with some fonts the labels C11 and Cl1 may look very similar, while C(11) and Cl(1) are clearly distinguished]. If the program cannot provide the text you need, you may be able to transfer the plot into a graphical manipulation program (e.g. Corel-Draw, Powerpoint or PhotoShop) or use rub-on lettering, but stencilling is no longer acceptable and most journals will refuse to transfer handwritten labels onto an unlabelled copy of the plot. It is not always necessary to label every last atom, but any atoms referred to in the text or in a table of selected molecular geometry parameters should be identified: for example, in a coordination compound, labelling the central metal and the ligand donor atoms may suffice. (You can always include a fully labelled version for the referees and possibly for deposition.) Unless they are of special significance (e.g. involved in H-bonding or transition metal hydride complexes) hydrogen atoms are not normally labelled.

Other text may be included: this can be excellent on posters, but where figures will undergo reduction it is likely to become hard to read. For example, adding bond lengths and angles to a drawing may help interpretation, but only if they are legible. To avoid cluttered drawings, some journals expressly forbid these additions and relegate such text to the figure caption.

15.6   Some hints for effective drawings

(a) Decide on the content: this is largely fixed for a single molecule, but there is much more choice for packing diagrams. In some cases the hydrogen atoms make it impossible to see the rest of the structure and can be omitted, although you may wish to include those on O or N atoms, for example. You can sometimes reduce clutter by devices such as drawing a single bond (in a different style) from the metal to the centroid of a coordinated benzene (or cyclopentadienyl) ring rather than the six (or five) bonds to the individual carbon atoms. Sometimes you need to omit peripheral groups, or show only the *ipso* carbon of an aryl ring, before you can see the salient parts of the molecule (you must acknowledge this in the figure caption). In some cases you may find it impossible to show all the important features in a single view.

(b) Invest some time looking for the best viewing direction with the minimum of overlap, especially where important atoms are concerned. If an atom simply cannot be manoeuvred into sight you could add a phrase such as 'C8 is wholly obscured by C7' to the figure caption.

(c) If the important features are still not obvious, can you emphasize them by using a distinct style for the atoms or bonds involved? For example, you can identify a metal's coordination sphere by having a distinct style for the ligand–metal bonds. If an atom has additional coordination at a greater distance, the bond(s) involved can be shown differently.

(d) If colour is effective, use it in moderation to draw attention to selected features of the structure.

(e) Choose the most effective representation to convey the information you want, bearing in mind that some journals may have specific requirements. Displacement ellipsoid plots certainly contain a lot of information, but it may not be the information you want to convey. Often the most significant atoms appear smallest because they have higher atomic and coordination numbers and consequently have lower displacement parameters. Furthermore, there is limited scope for differentiating atom types (but see Fig. 15.5), whereas a ball-and-spoke model allows more freedom to assign atomic radii and drawing styles. Avoid the use of similar styles for different atoms as far as possible: for example, styles 7, 8 and 9 in Fig. 15.3 may look identical if the circle representing the atom is very small (e.g. in a packing diagram) or after reduction.

(f) Avoid clutter. In some cases you have to add atom labels, but it may be possible to be selective. Omitting parentheses may help, as will calling the only phosphorus atom in the structure P rather than P1 or P01. Also, you may be able to label the carbon atoms using only their numbers (i.e. omitting the atom type and any parentheses). If there is no room to place a label close enough to an atom to identify it uniquely, consider placing the label some distance away with a line or arrow pointing to the atom.

(g) Take particular care with stereo views. Do they constitute a good view? Most importantly, are they better than a larger mono view occupying the same space?

(h) As mentioned earlier, different criteria apply for different publication formats. Are you preparing an illustration for a journal, a thesis, a poster, a Web page or an overhead transparency? Do not unthinkingly transfer a figure between formats without assessing its suitability. For example, a Web graphic that is so complex that it takes several minutes to download, or which is only effective on a very high resolution monitor, is unlikely to reach a wide audience.

(i) If you are submitting results to a journal that allows only a limited number of views (e.g. one) of any single structure, consider whether figures can be combined without loss of information.

(j) Be creative and have fun—this part of crystallography allows you more choice than any other. The original ORTEP manual [1] exhorted users to improve on the standard views produced from the program; that is now easier than ever.

15.7 Tables of results

The main tables produced at the end of the structure refinement will comprise all or most of the following:

- fractional atomic coordinates (with s.u.s; see Chapter 13 for a full treatment of these)—and possibly U_{eq} or B_{eq} values—for the non-H atoms: the values may be multiplied by a convenient factor (given in the table heading) to give integers, or expressed as decimal numbers;
- atomic displacement parameters—normally as U^{ij} or B^{ij}—with s.u.s;
- fractional atomic coordinates—and possibly U_{iso} or B_{iso} values—for H atoms which have not been refined freely (those which have been refined could either be given here, with s.u.s, or moved into the main table of fractional coordinates);
- molecular geometry parameters (bond lengths, valence angles, torsion angles, intermolecular contacts, least-squares mean plane data, etc.); there will usually be two versions of these tables, a shorter list of selected parameters for publication and a fuller listing (of the bonds and angles at least) for refereeing and deposition;
- structure factor tables.

It is also possible to tabulate crystal data and details of the structure determination, although this is not efficient in terms of space unless you can combine data for at least two or three structures in one table. Journals may have particular requirements, but if they are not specified (perhaps because the journal rarely publishes crystal structures) those of the journals of the Royal Society of Chemistry (www.rsc.org) or the American Chemical Society (pubs.acs.org) seem to be widely accepted.

Journals still vary enormously in their policies on crystallographic data—what they will publish, what they require as supplementary data, and what they will deposit. Before you start to prepare a submission, study the relevant instructions for authors (usually published in the first issue of each year) and follow them closely. There is, however, a strong trend towards publishing less, and many journals have stopped publishing structure factors, displacement parameters, fractional coordinates and full molecular geometry (more or less in that order chronologically), so that the selection of results for publication assumes greater importance (see below). Many journals will accept supplementary data in CIF format rather than hardcopy, although a few still seem a little suspicious of electronic data and demand both. Some refinement programs will produce tables of results automatically, and although these are useful they almost always benefit from critical inspection and sometimes manual adjustment of content and format, but you must exercise extreme care not to introduce numerical or other errors.

15.8 The content of tables

(a) Selected results
This almost always affects molecular geometry parameters, as coordinate data are usually complete—it is not permissible to include only the coordinates of what you consider the 'interesting' atoms! The selection of geometry parameters depends on the chemical nature of the compound and the structural features that you want to emphasize. These

are often obvious: in a coordination compound you might only want to include bonds involving a central metal and angles subtended at it (torsion angles involving such metals may be produced automatically but in most cases are not even worth archiving), but each structure should be considered individually. There is no point in trying to publish bond lengths that have been constrained during refinement, or that are unreliable because they fall in a region of the structure affected by disorder. Extensive listing of the internal molecular geometry of typical phenyl rings, whether constrained or not, is useful only for refereeing purposes and deposition. In many organic compounds there are no interesting or unusual bond lengths or angles that merit publication, but a selection of torsion angles might be worth including. Mean values or ranges may usefully supplant large numbers of individual values for similar parameters.

(b)   Redundant information

Where a molecule lies on a crystallographic symmetry element, some of its molecular geometry parameters will be equal or simply related to each other and therefore not all need to be given. Strictly speaking, only the unique set should be given and automatic table-generating routines may not be able to cope with this requirement. It may, however, be sensible to include some redundant information to make the situation clearer, especially for a non-crystallographic audience. For example, molecules of doubly-bridged dinuclear metal complexes $M_2(\mu\text{-L})_2$ contain four-membered rings and these are often found lying across crystallographic inversion centres: by symmetry, the opposite M–L bond lengths are equal; the two M–L–M angles are equal; the two L–M–L angles are equal; the MLML ring is strictly planar; and adjacent pairs of M–L–M and L–M–L angles add up to exactly 180°. The independent parameters are two adjacent M–L bond lengths and one angle within the ring. A mirror plane or a twofold rotation axis instead of an inversion centre will involve different relationships among the parameters, and these relationships will depend on the orientation of these symmetry elements as seen in Exercise 14.3. Similar arguments apply to the more common situation where a structure contains a central atom, often a metal, on a special position. For example, a four-coordinate palladium atom on an inversion centre has only two independent bond lengths and one independent angle.

When tables contain atoms that are related by symmetry to those in the original asymmetric unit, for example in order to give a bond length between two atoms related by a mirror plane, these atoms must be clearly identified (e.g. C5′, C5* and C5^i could be symmetry equivalents of atom C5) and the symmetry operations denoted by ′, * or ^{i} defined in a footnote.

(c)   Additional entries

Not all entries required for a molecular geometry table are necessarily produced automatically. 'Long' bonds may be missed and have to be inserted manually; short contacts such as those in hydrogen bonding may be calculated elsewhere but not transferred automatically. You may even want to include non-existent 'bonds', for example to demonstrate that two atoms are not close enough to interact. These values and their s.u.s should be calculated by the refinement program.

15.9   The format of tables

Journals tend to have their own requirements for tables, which must be followed—or you may have problems with the referees or editors. The precision to which results are required does vary: *Acta Crystallographica* prefers s.u.s in the range 2–19, while *Dalton Transactions* prefers 2–14, and some referees and editors object to s.u.s of 1. This can mean allowing more significant figures for the coordinates of heavier elements. Make sure the s.u.s look sensible and that any redundant data (such as the U^{ij} components for atoms on special positions other than inversion centres) have the correct relationships among their values (and among their s.u.s).

Ensure that the table headings are informative and correct: are the powers of ten quoted there actually those used in the table? Are the displacement parameters correctly identified as U or B? Are the headings on the structure factor tables correct, with any reflections not used in the refinement flagged? If it is possible, we suggest having a compound code or other identifier on every page of tables so that structures cannot be mixed up.

While one program might produce geometric parameters based on the order in which atoms occur in the refinement model, another might give the bonds in ascending order of length, and so on. It is worth looking at the tables to see whether this can be improved. In our opinion, the clarity of this table of selected bond lengths (in Å):

$$
\begin{array}{ll}
\text{Pd–N6 } 1.996(8) & \text{Pd–N2 } 2.017(7) \\
\text{Pd–N4 } 2.001(6) & \text{Pd–N5 } 2.035(6) \\
\text{Pd–N1 } 2.008(7) & \text{Pd–N3 } 2.057(7)
\end{array}
$$

is much improved by re-ordering to give:

$$
\begin{array}{ll}
\text{Pd–N1 } 2.008(7) & \text{Pd–N4 } 2.001(6) \\
\text{Pd–N2 } 2.017(7) & \text{Pd–N5 } 2.035(6) \\
\text{Pd–N3 } 2.057(7) & \text{Pd–N6 } 1.996(8)
\end{array}
$$

15.10   Hints on presentation

(a)   In research journals

This has mostly been covered already. Follow the instructions to authors for submitting papers, including experimental data, tables, figures and supplementary data. What is the policy on colour plots? The range of formats for literature references can appear overwhelming, and if you submit to a wide range of journals, the use of reference management software may be worthwhile. If you regularly use the same references in the same format they could simply be stored in a standard ASCII or word processor file.

(b)   In theses and reports

Here you have much more freedom, but you have to be careful that the result is appropriate in style and length to the purpose in hand: a thesis which runs to 300 pages may be acceptable, but an interim report of that size is ridiculous. Fortunately, guidelines are normally available, so consult them before you write a word. Don't overdo the tables: it is seldom necessary to include structure factor tables, even in an appendix. However, you could put coordinate data and full molecular geometry tables in appendices and retain only selected information in the main body: this will cause less disruption to the flow of your report. You can be more generous with diagrams than when publishing in a journal, but remember that there must be a good reason for including any diagram. Do your local rules allow appendices to be submitted on microfiche or CD?

(c)   On posters

Select the most important points you want to get across. Save yourself time by planning what you are going to present—there is no point in producing material you don't have space for. (Do you know the size and orientation of the poster display area available to you?) Text must be readable from a distance of one to two metres—you may not always attract a crowd, but the poster will make a greater impact if it can be viewed from a comfortable distance. Keep any tabulated information short and relevant. (Are the crystal data really needed on the poster, or is it sufficient to have these to hand in case someone asks?) Posters are one place where colour can be exploited to the full, and not only in figures but in text, backing material and surrounds. It is harder to overdo it here, but still possible! You should keep the information density relatively low: this has the additional advantage that you will have something more to tell those who express an interest in your work.

(d)   As oral presentations

Many of the points mentioned in respect of posters apply here too: avoid high-density slides or overhead transparencies which nobody will have time to read. Don't be tempted to use that convenient table prepared for publication if it consists mostly of values which are not relevant to your lecture.

An important function of your visual aids is to remind you what to say next, so they must be 'in phase' with your talk; they must not let the audience see your final results while you are still outlining the problem! If you need to refer to a slide or transparency at two points in your talk, it is better to make two copies than to waste time rummaging around for the single copy you last saw ten minutes before. If you have just covered the material outlined on a slide and made a point which requires the audience to absorb information from the slide, do not immediately snatch it away and replace it with the next one. This point is particularly relevant if the slide contains a crystal structure diagram.

Compose your visual aids carefully. Try to find out the size of the auditorium and about its facilities. Mixing slides and transparencies requires some planning to ensure you don't lose track of what you are saying. If you are inexperienced it is safer to have all your visuals in the same format if at all possible. (Using video projectors or other advanced facilities requires planning and practice, even for experienced lecturers.) Colour can be extremely powerful, especially when used in bold, simple illustrations.

Do not try to cover too much material. Your audience will be less familiar with your material than you are and it will not help if you speak too quickly. Time your talk in advance—a practice session with a sympathetic (and constructively critical) audience is a good idea if you are unused to speaking in public. You should assess the composition of your audience in advance. If they are not experts in your own field you will need to give more background information so that they understand the context, before you begin to describe your own work and its results.

Humour can be a good way to engage the audience's attention but it needs to be used carefully and sparingly. In some circumstances it is wholly inappropriate. If in any doubt, avoid it.

(e) On the Web

The Web is an excellent medium for disseminating research results, but it has its own special requirements. The speed at which data can be transferred is limited: as faster networks are installed these are required to carry ever more demanding applications. Combined with a continuing, almost exponential, increase in the number of users, this ensures that bandwidth is always restricted. As a result, the most effective Web pages are often those which do not involve large-scale data transfer in order to be useful. Authors need to bear in mind that many of their potential audience may be using modest hardware and not-so-recent software, and they must ensure that using new features in the latest version of their Web authoring software does not restrict this audience. On a related theme, any Web page should at the very least be viewable with the two most common Web browsers (Netscape Navigator and Internet Explorer).

It might seem obvious that any website should be designed so that the visitor can find information easily, yet many impressive-looking corporate sites are so poorly structured that finding what you want is time-consuming, inefficient and frustrating. If you have an extensive website you should give serious thought to how it is constructed, and in particular whether your home-page allows a visitor to begin navigating it easily.

As well as hyperlinked text and graphics, Web publishing offers the possibility of illustrations which can be rotated or otherwise moved, either according to your preprogrammed instructions or in response to user input. This allows the visualization of complex structures and packing diagrams, for example. These features are implemented by means of Virtual Reality Modelling Language (VRML) extensions to your browser.

Publishing results on the Web is quite different from displaying them on a poster that you can take down at the end of a conference. Once placed on the Web, material assumes a substantial degree of permanence as it is accessed, stored in various caches, copied to other formats and printed. On the other hand, such results can be more ephemeral that those printed in a journal, as it is possible to remove, modify or update them. The copyright implications of publishing your results on the Web are often far from clear, but these are likely to become more serious with the spread of electronic publishing. As with other forms of publication, placing results on the Web should only be done with the knowledge and consent of all those contributing to the work, and only when the consequences of doing so have been fully explored.

15.11 Archiving of results

Although some of the results of your structure determinations will be deposited in a database after publication, you must keep your own copies of all relevant files and other information safely and in an accessible form. Most crystallographers know the frustration of setting out to prepare a structure for publication, only to find that some experimental parameter such as the colour of the crystal or the type of diffractometer used is not immediately available and has to be ferreted out. In the not-so-distant past the only safe way seemed to be to keep every piece of hardcopy output ever generated for a structure, but now archiving and transmission tools such as the CIF format allow this to be done much more concisely. The 'paperless office' promised by the proponents of information technology a decade ago may have proved illusory elsewhere, but in the modern crystallography laboratory it has largely materialized. The CIF and its uses are described in detail in Chapter 16.

When archiving data the main considerations are safety and accessibility. To address the first, you need to keep backup copies of your files, possibly on tape, and including a set which will survive fire, flood and theft at your workplace. This could involve a fireproof safe, but keeping a backup at home is probably as reliable. Keeping all the files for one structure together aids organization, and utilities such as PKZIP and its various Windows variants allow these to be compressed within a single archive file, with the bonus of a considerable saving on disk space. For accessibility, you need some form of indexing so that the structures you require can be quickly and uniquely identified. Before starting to rely on it, you must check the backup procedure works by restoring some typical files from the archive. Most backup devices come not only with software to drive them but also with documentation which includes advice on how to implement a suitable and effective backup regime. This documentation should include explanations of how to use both full and incremental backups, the latter referring to the procedure whereby only those files which have changed since the last backup are transferred to the archive.

There are two distinct aspects to backing up a particular computer. The first requires an archive medium capacious enough to allow you to back up everything on that computer, including operating system, applications and data. This would allow you to recreate your working environment in the event of the computer or its hard disk failing totally, and requires a backup medium with the same capacity as your hard disk. The best solution is probably some kind of tape drive, and models with capacities of up to several tens of Gbytes are currently available. An additional level of security could be provided by backing up your working hard disk to another *physical* hard disk (not a different *logical* drive on the same disk) on the computer. Such a backup will survive any failure of the working disk and most disasters short of theft or outright destruction of your computer. If the computer is automatically backed up over a network you may be content to rely on this, but make sure that the frequency of backup is appropriate and that the backup files are accessible. Remember to update the backup files whenever you make significant changes to the computer, for example after a major new application has been installed and configured.

The second aspect is the regular backup of new data, and the media that can be used will depend on the volume of these data. The essential files to be kept at the end of a structure analysis may well amount to only a few hundred kilobytes after compression and several structures could be archived on a standard 3.5-inch floppy disk. The same may well be true for the important files from a data collection on a four-circle diffractometer, but the frames for one data collection using an area detector diffractometer occupy several hundred Mbytes and archiving to CD-ROM is a sensible option.

When you are planning a backup regime an important factor is ease of use. Any method which is cumbersome or time-consuming is unlikely to encourage you to carry it out regularly. Archive media can change and develop as rapidly as other aspects of computer hardware and factors such as capacity, cost, convenience and durability need to be considered. It is not necessarily best to adopt the latest technology; in fact it may be safer to select one that has gained reasonably wide acceptance so that consumables such as tapes are likely to remain available for the useful life of the computer. The following table gives (sometimes approximate) capacities for various storage media:

Medium	Capacity
3.5-inch disk	1.44 Mb
Super-floppy	120 Mb
Cartridge	100 Mb–20 Gb
CD-ROM	550–650 Mb
DVD	up to 5 Gb
Tape	400 Mb–70 Gb

References

[1] C. K. Johnson, *ORTEP*, Report ORNL-3794, Oak Ridge National Laboratory, Oak Ridge, Tennessee, 1965.

[2] C. K. Johnson, *ORTEP*II, Report ORNL-5138, Oak Ridge National Laboratory, Oak Ridge, Tennessee, 1976.

[3] G. M. Sheldrick, *SHELXTL XP* graphics module: various versions, and for different platforms (e.g. PC, SGI), Bruker AXS Inc., Madison, Wisconsin, USA.

[4] L. J. Pearce, D. J. Watkin and C. K. Prout, *CAMERON*, University of Oxford, 1994.

[5] S. R. Hall and D. du Boulay (eds), *Xtal_GX*, University of Western Australia, 1995.

[6] L. J. Farrugia, WinGX—A Windows program for crystal structure analysis, University of Glasgow, Scotland, 2001.

[7] E. Keller, *J. Appl. Cryst.*, 1989, **22**, 19.

[8] R. X. Fischer, *J. Appl. Cryst.*, 1985, **18**, 258.

Exercises

15.1   What possible problems are indicated by the displacement ellipsoid plots in Fig. 15.11?

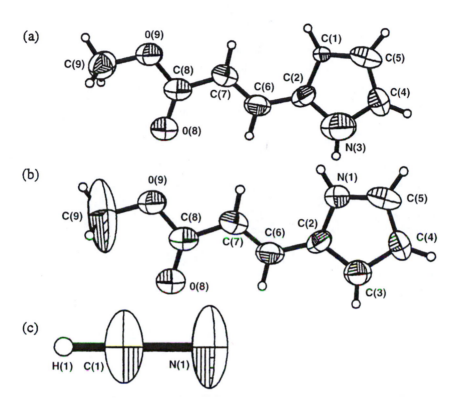

Fig. 15.11.   Ellipsoid plots for Exercise 15.1.

15.2   Prepare a caption for Fig. 15.8. What could be added to the figure to increase its information content? How would this addition affect the caption?

15.3   The following table was submitted for publication. How could it be improved?

Table 2.

Cd4 O4	2.102(9)	Cd4 S4 2.278(4)	
Cd4 S2	2.281(4)	Cd2 Cl2	2.381(4)
Cd4 Cd1	4.204(1)	Cd3 O3	2.125(9)
Cd3 S3	2.258(4)	Cd3 S4 2.307(4)	
Cd3 Cl3	2.364(4)	Cd3 Cd2	4.171(1)
Cd2 O2	2.073(11)	Cd2 S2 2.265(4)	
Cd2 S1	2.306(4)	Cd4 Cl4	2.342(4)
Cd1 O1	2.182(8)	Cd1 S1 2.263(4)	
Cd1 S3	2.294(4)	Cd1 Cl1	2.345(4)

15.4   How could the following table be improved before inclusion in a thesis?

Table 1.   Crystal data and structure refinement for 1 at 150(2) K.

Empirical formula   C24 H33 N3
Formula weight   363.53
Crystal description   colourless tablet
Crystal size $0.54 \times 0.50 \times 0.27$ mm
Crystal system   Monoclinic
Space group   P2(1)/c
Unit cell dimensions   a = 12.736(6) A   alpha = 90 deg.
                                           b = 15.371(6) A   beta = 94.12(8) deg.
                                           c = 10.676(5) A   gamma = 90 deg.
Volume   2084.6(16) A^3
Reflections for cell refinement   58
Range in theta   12.5 to 17.5 deg.
Z   4
Density (calculated)   1.158 Mg/m^3
Absorption coefficient   0.068 mm^−1
F(000)   792
Diffractometer type   CAD4
Wavelength   0.71073 A
Scan type   omega/theta
Reflections collected   3680
Theta range for data collection   2.65 to 25.03 deg.
Index ranges $-14 <= h <= 15, 0 <= k <= 18, 0 <= l <= 12$
Independent reflections   3680 [R(int) = 0.000]
Observed reflections   2412 [I > 2sigma(I)]
Decay correction variation   +/−7%%
Structure solution by   direct methods
Hydrogen atom location   calc
Hydrogen atom treatment   constr
Data/restraints/parameters   3680/0/253 (least-squares on F^2)
Final R indices [I > 2sigma(I)]   R1 = 0.0585, wR2 = 0.1304
Final R indices (all data)   R1 = 0.0975, wR2 = 0.1578
Goodness of fit on F^2   1.039
Final maximum delta/sigma   0.000
Weighting scheme
calc w = 1/[\s^2^(Fo^2^) +(0.6200P)^2^+0.0P] where P = (Fo^2^+2 Fc^2^)/3
Largest diff. peak and hole   0.21 and −0.21 e.A^−3

16

The Crystallographic
Information File (CIF)

16.1 Introduction

The Crystallographic Information File (CIF) is an archive file for the transmission of crystallographic data: this transfer can be between different laboratories or computer programs, or to a journal or database. The file is free-format, flexible and designed to be read by both computer programs and humans (the latter require a little practice at the start). The specification of the CIF standard has been published and the same article provides information on its evolution [1]. It is based around the Self-defining Text Archive and Retrieval (STAR) procedure [2], and consists of data names and the corresponding data items, with a loop facility to handle repeated items such as the author/address list or the fractional coordinates. The format is extensible, so that data names covering new developments such as area detectors can easily be accommodated. However, once a data name is included in the recognized dictionary, it is never removed, otherwise portions of those archives written in the interim would be undefined.

16.2 Basics

The CIF is an ASCII file, such that only the following characters are allowed:

abcdefghijklmnopqrstuvwxyz

ABCDEFGHIJKLMNOPQRSTUVWXYZ

0123456789 !@#$%^& * ()_ + {} : "~⟨⟩?|\- = []; `, /.

Any others which you may want to include in a manuscript (such as Å,°, é, ø, subscripts and superscripts, Greek letters, mathematical symbols such as $\pm$, $\geq$ or ∞, and chemical multiple bonds) require special codes which are detailed in the *Notes for Authors* for *Acta Crystallographica, Section C* [3]. Do not attempt to show italicized, bold or underlined text (e.g. in space group symbols), as these attributes should be added automatically. Many data names have implicit units (e.g. Å for _cell_length_a; Å^3 for _cell_volume; minutes for _diffrn_standards_interval_time) and these units must not be appended; thus

_cell_volume   2367.5(8)

is correct but

_cell_volume   2367.5(8) \%A^3^

is not.

If you prepare your CIF using a word processor rather than a text editor, you must make sure that the output file is ASCII, that there are no (hidden?) embedded codes, and that lines do not exceed 80 characters. If you are exporting an ASCII file from a word processor it is best first to select a large fixed font (e.g. Courier 12 point or larger), so that any lines which are part of blocks of text do not overrun upon conversion. Do not include any non-ASCII characters in the word processor file, as these will probably be lost or corrupted upon writing the ASCII file, and even if they survive unchanged, the CIF processing software will not recognize them correctly.

In 2001 the Cambridge Crystallographic Data Centre (CCDC) developed a free CIF editor (enCIFer), which allows safe and easy editing of the CIF produced by a refinement program. This editor aims to provide an intuitive interface to allow manipulation of CIFs without the risk of compromising its strict syntax. An editing wizard assists the generation of a CIF with all the necessary data, and entries are checked for syntax and validated before being entered into the CIF; template files can be used to minimize unnecessary retyping; checks are performed to identify any missing data items; and data loops can be edited via a spreadsheet-style window. To assist the user, on-line help is available for data items, structures can be visualized in 3D and as a 2D chemical diagram. Using a web-style browser, it is possible to load in a number of separate CIFs (which could be global blocks, templates, or output from a refinement program), which are then represented by their icons: dragging and dropping these icons allows the user to combine the CIFs as required. Unix and Windows versions of enCIFer can be freely downloaded from the CCDC web site (www.ccdc.cam.ac.uk).

The following CIF terminology is used:

text string a string of characters delimited by blanks, by quotes, or by semicolons (;) as the first character on the line.

data name a text string starting with an underline (_) character.

data item a text string not starting with an underline, but preceded by a data name

data loop a list of data names, preceded by loop_ and followed by a repeated list of data items.

data block a collection of data names and data items (which may be looped) preceded by a data_code statement and terminated by another data_ statement or the end of the file. A data name may only occur once within any one data block.

data file a collection of data blocks: no two data block codes may have the same name.

CIF data name and data block code definitions are restricted to a maximum of 76 characters, but their hierarchical construction and careful design mean that they are largely self-explanatory: e.g. _publ_author_name, _exptl_crystal_colour and _computing_structure_solution.

The term 'CIF' has acquired a range of informal meanings: it is used to describe the format, the data output by a control or refinement program, as well as the data file (comprising two or more data blocks) submitted as a manuscript for electronic publication.

16.3   Uses of CIF

(a) Your own local archive. A CIF produced by a refinement program upon convergence of your structure can be edited and augmented so that all the relevant results and details of procedures can be stored. As with other uses, the degree of manual editing required will depend on the extent to which your data collection and reduction programs produce relevant output in CIF format.

(b) A standard method of transmitting data between crystallographic programs (an increasing number of which read files in CIF format) or to colleagues in other laboratories.

(c) An efficient method of providing supplementary data for papers containing crystal structure determinations.

(d) A standard route for deposition into structural databases.

(e) A route to standard printed tables (e.g. via the SHELXL ancillary program CIFTAB).

(f) Direct electronic submission of manuscripts to journals such as *Acta Crystallographica* or *Zeitschrift für Kristallographie*. The required data file will consist of a number of data blocks. One, perhaps called data_global and possibly prepared by manual editing of a template, will contain author contact information, submission information, an author/affiliation/address list, a title, a synopsis, an abstract, a comment (discussion) section, experimental details, references, figure captions and acknowledgements. This block will be followed by one data block for each structure to be included, perhaps called data_compound1, data_compound2, etc.

16.4   Some properties of the CIF format

(a) Within any data block, the ordering of the associated pairs of data names and data items is not important, as file integrity does not depend on finding these in a particular sequence. However, human readers will find it easier to read the CIF if they are grouped logically. Furthermore, there are no restrictions on the ordering of the data blocks.

(b) Each data name must have its corresponding data item, but the latter need not contain real information. Sometimes a placeholder such as '?' or '.' is used, as in
_chemical_name_common ?
These placeholders are used within loop structures when some data items are not relevant to every line of the loop. In the following example the fourth data name in the loop only applies in the second line of the looped data items.

```
loop_
_geom_bond_atom_site_label_1
_geom_bond_atom_site_label_2
_geom_bond_distance
_geom_bond_site_symmetry_2
_geom_bond_publ_flag
Ni N1 2.036(2) .      Yes
Ni N1 2.054(2) 2555 Yes
Ni  S2   2.421(10) . Yes
```

```
C   S2   1.637(3)  . Yes
N1  C1   1.327(3)  . ?
N1  C5   1.358(4)  . ?
N2  C12  1.309(3)  . ?
```

Some items are mandatory for certain CIF applications: for example, the list of data items required for a submission to *Acta Crystallographica, Section C* is given in that journal's *Notes for Authors* [3].

(c) Certain data items can be specified as standard codes, and these must be used wherever possible. For example, there are now eight standard codes for the treatment of H atoms during refinement associated with the data name _refine_ls_hydrogen_treatment and these include refall (all H parameters refined) and constr (e.g. a riding model). If the standard codes are inappropriate or inadequate, then a fuller explanation can be given as part of an experimental section.

(d) There are now a large number of ancillary programs available for preparing, check-ing, manipulating and extracting CIF data. Some of these will be mentioned later and others are given on the IUCr website (www.iucr.org).

(e) Additional tables can be created within the CIF format. The most common of these contain hydrogen-bonding parameters and there are now standard _geom_hbond_ data items to facilitate their input. However, it is possible to set up tables of non-standard parameters by defining additional data items: the following example lists short Si··· O intermolecular contacts:

```
loop_
_publ_manuscript_incl_extra_item
_publ_manuscript_incl_extra_defn
'_geom_extra_table_head_A'          No
'_geom_table_footnote_A'            No
'_geom_bond_atom_site_label_Si'     No
'_geom_bond_atom_site_label_O'      No
'_geom_contact_distance_SiO'        No
'_geom_contact_site_symmetry_O'     No


_geom_extra_table_head_A
;
Intermolecular distances for contacts of the type Si··· O (\%A)
;


loop_
_geom_bond_atom_site_label_Si
_geom_bond_atom_site_label_O
_geom_contact_distance_SiO
_geom_contact_site_symmetry_O
    'Atom Si' 'Atom O' Distance 'Symmetry operator^a^'
        Si1   O6       2.363(4)  1555
        Si3   O2       2.420(6)  1545
        Si4   O7       2.227(3)  2767
```

_geom_table_footnote_A

;

^a^ Applies to oxygen atom in each case

;

16.5   Some practicalities

16.5.1   *Strings*

The correct handling of strings is vital. There are three ways to supply the information in these, and examples of each follow.

(a) Delimitation by blanks: the data item is effectively a word or a number without spaces within it. The data item cannot extend beyond the end of a line. Examples are:

```
_publ_contact_author_email       J.O-Groats@north.ac.uk
_cell_length_a                   10.446(3)
_diffrn_standards_number         3
```

Note that J. O-Groats@north.ac.uk and 10.446 (3) are not allowed.

(b) Delimitation by quotation marks: the data item may now contain spaces. It is limited to one line, but it can be on the line following the data name if required. For example:

```
_exptl_crystal_density_method   'not measured'
_chemical_formula_moiety   'C12 H24 S6 Cu 2+, 2(P F6 −)'
_publ_section_acknowledgements
   "We thank EPSRC for a postdoctoral award (to J.O'G.)."
```

(c) Delimitation by semicolons as the first character in a line: this is necessary for blocks of text which exceed one line in length. For example:
```
_publ_section_abstract
```

;

In the title compound, C~12~H~22~O~7~, molecules occur exclusively as the cis geometric isomer and are linked by hydrogen bonding to form helices running parallel to the crystallographic c direction.

;

16.5.2   *Text*

(a) When preparing a data file for publication, most effort will be devoted to the textual sections, in particular _publ_section_abstract, _publ_section_comment and _publ_section_references. Much of this may appear as normal text, but certain special character codes are commonly required. Subscripted and superscripted text is delimited by pairs of tilde ($\sim$) and caret (^) characters, respectively: for example, if you want $[Cu(H_2O)_4]^{2+}$ to appear in your paper, you need to enter it as [Cu(H~2~O)~4~]^2+^. Note that these must not be used in _chemical _formula_moiety, etc. In discussing molecular geometry you will need the symbols

Å and °, the codes for which are \%A and \% (not ^o^), respectively. You will occasionally need other codes; see the appropriate page in reference [3] for the full list.

(b) Certain trivial errors occur frequently and can cause a great deal of annoyance, because the CIF processing software does exactly what you tell it to, rather than what you want. CIF checking software (see below) will strive to return helpful reports on the location of errors in the data file, but sometimes the results of the fault are so pervasive, or appear so far removed from the original error, that a manual search is necessary. Some of the most common and irritating faults arise from the simplest of causes, such as the failure to have matching subscript and superscript codes: forgetting to 'switch off' these features means that subsequent information in the CIF is misinterpreted. Another frequent mistake is not terminating text strings correctly. By using enCIFer, you should be able to avoid these problems.

(c) Within a CIF, the selection of molecular geometry parameters for publication depends on the setting of the _geom_type_publ_flag for each parameter, where type is bond, angle or torsion. Setting a flag to Yes (or y) indicates that the corresponding parameter should be published: anything else (No, n or ?, for example) means that it will not. Make sure that this editing does not disrupt the number of data items (including placeholders), as doing so will create problems for any program attempting to read the CIF.

16.5.3   Checking the CIF

Before a CIF is used, for whatever purpose, it is highly advisable to submit it for automatic checking. This may be done by sending the CIF by email to **checkcif@iucr.org** and inspecting the report which is returned. The **checkcif** procedure tests for valid CIF data names, correct syntax, missing IUCr Journals Commission requirements, consistency of crystal data, correct space group, unusual atomic displacement parameter values, completeness of diffraction data, etc. Depending on the intended purpose of the CIF, failure to satisfy certain of these tests (e.g. missing Commission requirements) may not be important, as these are designed primarily as a check on a data file before electronic submission to *Acta Crystallographica*. If you have access to a Postscript printer, or a viewer such as ghostview, you should take advantage of another utility by emailing your CIF to **printcif@iucr.org**. A preprint of your paper will be returned for checking: this is valuable because some errors, especially formatting ones, may not be detected by **checkcif** but they are usually horribly obvious on the preprint. The **checkcif** and **printcif** facilities are also available through the IUCr website (www.iucr.org). By this route **printcif** offers the additional options of returning your preprint as a PDF file or PDF stream. You can also carry out useful additional checks with programs such as PLATON [4] running on your own computer.

References

[1] S. R. Hall, F. H. Allen and I. D. Brown, *Acta Cryst.*, 1991, **A47**, 655. Reprints are available from the International Union of Crystallography, 5 Abbey Square, Chester CH1 2HU, UK.

[2] S. R. Hall, *J. Chem. Inf. Comput. Sci.*, 1991, **31**, 326.

[3] *Acta Cryst.*, 2001, **C57**, 129–136.

[4] PLATON, a program for the automated analysis of molecular geometry. A. L. Spek, University of Utrecht, The Netherlands.

Exercises

16.1 This is a partial version of a CIF that was submitted for publication. The errors encompass many of the mistakes commonly made when preparing a CIF for submission. Identify as many of them as you can.

```
ü©ü©ü©ü©ü©ü©ü©ü©ü©# ===============================
     INITIAL PUBLICATION REQUEST
# =============================================
_publ_contact_author
'Prof. John Smith'
;
Faculty of Pharmaceutical Sciences
John Smith University
John Smithsville
IN 40678
U.S.A.
;
_publ_contact_author_email      J. Smith@xray.jsu.edu
_publ_contact_author_fax        '(1) 727 564 2209'
_publ_contact_author_phone      '(1) 727 564 2925'
_publ_contact_letter
;
_publ_requested_coeditor_name   ?
_publ_requested_journal         'Acta Crystallographica, Secti

_publ_author_name
_publ_author_address
'John Smith'                    # lastname, firstname
;
Faculty of Pharmaceutical Sciences
John Smith University
John Smithsville
IN 40678
U.S.A.
;
# =============================================
# TEXT
# =============================================
  _publ_section_title
; A new copper(àá) complex of acetic acid
;
_publ_section_abstract
; In the title compound, bis(acetato)copper(àá)dihydate
```

there exists a three-dimensional network of N–HücO and O–HücO
intermolecular hydrogen bonds.

; _publ_section_comment
; Acetic acid, (àf),and its derivatives have been the subject of recent interest. For
example, copper (àá) dihydrate, (àá) has been determined as a model compound for
probing the interaction ....

....
The crystal structure is stabilized by the N–HücO and O–HücO
hydrogen bonds involving the water molecules:
N(3)–H(3)ücO(2^i^)=2.82(5);N(1)–H(1)ücO(3^ii)=2.82(5);
O(3)–H(31)ücO(4^iii^)=2.95);

....
;

===
EXPERIMENTAL
===
_chemical_compound_source
; The green acicular crystal was obtained by the slow
concentration of an aqueous solution at room temperature.
;
_chemical_formula_sum	'C9 H6D6N6O12 Cu'
_chemical_formula_moiety	Cu(II)[(CH3N3O4)~4~(D2O)~2~]2(D2O)'
_chemical formula_weight	'444.44'
_symmetry_cell_setting	'triclinic'
_symmetry_space_group_name_H–M	'P1^-^'
_cell_length_a	5.267(2)
_cell_length_b	7.440(13)
_cell_length_c	9.872(2)
_cell_angle_alpha	86.15(2)
_cell_angle_beta	72.33(2)
_cell_angle_gamma	86.89(2)
_cell_volume	354.3 (4)
_cell_formula_units_z	1
_exptl_crystal_density_diffrn	'2.243'
_exptl_crystal_density_meas	?
_exptl_crystal_density_meas	'none'

....

_computing_data_reduction
'TEXSAN. TEXRAY Structure Analysis Package (Molecular Srtucture Corporation,
1985)'
_publ_section_references
;
Beurskens, P. T. (1984). Technical Report 1984, Crystallography lab. Toernooiveld,
6525 Ed Nijmegen, Netherlands.

....
Sheldrick, G. M. (1986). SHELXS86. Program for the solution of crystal structures,
University of G\"ottingen.

_publ_section_figure_captions
'Fig. 1'
; ORTEPII(Johnson, 1976) drawing of the title compound with the atomic
numbering scheme, viewed along the a axis.
Ellipsoids for non-H atoms correspond to 50% probability.
;
'Fig. 2'
; Packing diagram of the title compound along the a axis of the unit cell;
intermolecular hydrogen bonds are represented by dashed lines.
;

_publ_sectiom_table_legends
'Table I'
; Fractional atomic coordinates and equivalent isotropic thermal parameters
(\%A^2^).
;
'Table II'
; Selected bond length (\%A) and angles (^o^).
....

loop_
_geom_bond_atom_site_label_1
_geom_bond_atom_site_label_2
_geom_bond_distance
_geom_bond_site_symmetry_1
_geom_bond_site_symmetry_2
_geom_bond_publ_flag
Cu(1)   O(1)   2.446(8)   .   .   yes
Cu(1)   O(6)   2.468(5)   .   .   yes
Cu(1)   N(5)   1.985(4)   .   .   yes
O(1)   C(2)   1.241(7)   .   yes
C(1)   C(2)   1.394(9)   .   .   yes
....

; Anisotropic_displacement_parameters.
loop_
_atom_site_label
_atom_site_U~11~
_atom_site_U~22~
_atom_site_U~33~
_atom_site_U~12~
_atom_site_U~13~
_atom_site_U~23~
Cu(1) 0.023(4) 0.021(4) 0.017 (5) −0.007(3) 0.008(4) −0.016(5)
....
....

17

Crystallographic databases

Computer databases are collections of items of information with a common structure and format. They have a number of advantages over paper-based or other forms of information storage and retrieval. These include the greater ease of maintaining and updating them, the possibility of automatic validation of items as they are added, fast and powerful ways of searching them for items of interest (assuming appropriate computer software is available), compact storage and simple distribution, and the availability of individual selected items in electronic form for further analysis.

Any computer database has two components: the database contents themselves, and software for carrying out searches and analysis of the data.

Databases are important in many areas of chemistry and some are very well known. They include bibliographic databases, such as computer versions of the Science Citation Index and Chemical Abstracts; databases of commercial chemical catalogues and safety information; a range of NMR, IR and other spectroscopic databases; collections of thermodynamic data; and databases of chemical reactions for use in computer-aided synthesis.

They are particularly well suited for crystallographic applications, because of the systematic nature of crystallographic structural information. There are five main structural databases in current use internationally. They are maintained and developed by different organizations, and distributed to individuals and to national and regional computing centres, so they may be accessed either locally on a desktop computer or via network connections. Search software is tailored to the specific database; in some cases it is through a web browser interface.

17.1 Available structural databases

The Metals Crystallographic Data File (CRYSTMET, also known as MCDF or MDF) is maintained by the National Research Council of Canada. It contains information on metals and alloys, and also includes some binary compounds such as hydrides and oxides. This might seem to be a rather restricted topic, but in fact there were over 64,000 entries by the year 2001. Each entry has both bibliographic (authors, reference, compound or mineral name, formula) and numerical data (cell parameters, space group, atomic coordinates and displacement parameters if these are known).

The Inorganic Crystal Structure Database (ICSD) is the responsibility of the Gmelin Institut and the Fachinformationszentrum in Karlsruhe, Germany; it was initially developed at the University of Bonn. It consists of structures of inorganic materials and minerals, containing no organic carbon, because it was established specifically to complement the older Cambridge Structural Database; the definition dividing these two databases is a little arbitrary and there is a small degree of overlap in their contents. The majority of the structures in the ICSD are non-molecular, and chemical connectivity is much more difficult to define and assess for such compounds than for covalently bonded molecules. This makes the procedures for searching the ICSD generally less flexible and powerful, but one of the methods of searching is via a familiar style of web interface. The information provided is similar to that for MCDF. There are many closely related structures including isomorphous compounds, solid solutions and polytypes, and also multiple determinations of structures under different conditions, using different radiations, or by different research groups. By 2001 there were nearly 60,000 entries.

The largest, most developed and most widely used database is the Cambridge Structural Database (CSD), owned and administered by the Cambridge Crystallographic Data Centre (CCDC). Its compass is the whole range of organic, organometallic and metal coordination compounds broadly understood as molecular species, but excluding biological macromolecules. With about 240,000 entries by 2001, its growth continues to accelerate as modern developments in crystallography lead to more rapid data collection and structure solution. As well as bibliographic and numerical data similar to ICSD (but without displacement parameters), each entry is coded with molecular connectivity, allowing powerful search facilities based on structural fragments.

Biological macromolecules are provided for by the Protein Data Bank (PDB), which contains nucleic acids as well as proteins. It was developed at the Brookhaven National Laboratory, USA, but has recently been acquired by the Research Collaboratory for Structural Bioinformatics. Although small in terms of number of entries, it is also the fastest growing of the structural databases, with 14,000 entries by 2001, double the number of three years earlier. A separate companion database for nucleic acids NDB now exists, with about 800 entries.

The fifth database is rather different: the Crystal Data Identification File (CDIF), maintained by the US National Institute of Science and Technology. This stores unit cell and space group information, even when the full structure is not known, and is a very useful resource for checking unit cells obtained on a diffractometer in order to avoid wasting time collecting data for a known structure (such as a starting material or unexpected product of a reaction); see Chapter 5. By 1999 there were over 230,000 entries, with a backlog of new ones waiting to be processed.

In the UK most of these databases are available free of charge to academic researchers through the EPSRC-funded Chemical Databank Service at Daresbury Laboratory (http://cds.dl.ac.uk/cds/cds.html). Similar arrangements for access to some or all of the databases exist in other countries.

17.2 Contents of the Cambridge Structural Database

The five crystallographic databases have some features in common, but each has its own special properties. Since the CSD is the most widely used in chemical crystallography, we shall concentrate on some aspects of its nature and use.

Each structure in the CSD is identified by a unique reference code (REFCODE) assigned when it is added to the database. The REFCODE is a sequence of six letters which generally have no specific meaning; two numerical digits may be added in some cases to denote a second or further experimental study of the same material or a different crystalline form (a polymorph or different solvate); it acts as a convenient label.

The information for each structure is described by the CSD's developers as being under three headings. Three-dimensional data are the unit cell parameters, space group and atom coordinates, from which detailed molecular and intermolecular geometry can be calculated. Two-dimensional information is the stored chemical connectivity, showing which atoms are bonded to each other, together with atom types and charges, codes for different types of bonds, and a diagram of the chemical structural formula showing this connectivity; polymeric structures are flagged. All other information is referred to as 'one-dimensional': it includes all the bibliographic data, together with single numerical items such as temperature, calculated density, R factor; textual items such as compound name and colour; and comments on features such as disorder and errors in the published data.

The vast majority of entries in the CSD come from published literature, involving interaction with a large number of journals. Formerly, each entry had to be entered by keyboard and the connectivity diagram generated manually. Computer scanners made this easier but were by no means error-free. More recently, the introduction and widespread adoption of the CIF has led to increasing automation of the process. Some journals forward CIF submissions to CCDC, others require authors to do so before submitting manuscripts for publications, while others have no definite procedure. It is also possible for researchers to deposit structural results direct with CCDC for inclusion in the CSD without formal publication in a journal; the database is itself a form of publication, of course. All entries are thoroughly checked for consistency, for example by calculating molecular geometry from the coordinates and cell parameters and checking them against the geometrical parameters provided in the publication or CIF, and by comparisons such as the stated molecular formula with the list of atoms. Any errors are corrected if possible, or are flagged, and the database entry is, in some cases, more accurate than the original publication.

17.3   Searching the CSD

All of the items stored for each entry are available as bases for searching. Thus it is possible to find entries based on bibliographic or chemical information, on individual crystallographic data such as cell parameters, experimental conditions, precision of results, the presence of disorder, and many more. Particularly useful is the possibility of searching for structures containing a specified fragment that is constructed qualitatively in a drawing area (with a wide palette of templates and fragment groups available to facilitate this construction). Limits can be set on geometrical parameters such as distances and angles, in order to include or exclude structures of various kinds, and lists of search results can be manually filtered and refined.

From the retrieved structures (referred to as 'hits'), which may be few or many for any given search or combination of searches, simple lists may be generated, geometry may be calculated, diagrams can be obtained, the molecular structures may be manipulated on screen, and simple or more complex statistical analyses can be made of the results.

Until recently the CSD was available primarily on UNIX-based computing systems. There is, however, now an enhanced version of the search software that can also run under various versions of Windows on PCs. The current version of the software, together with the database, requires somewhat more than 500 Mbytes of storage, no problem on readily available computers.

There are many uses for the databases. The simplest is checking for the structure of a particular compound, in order to find out whether it has already been determined, but this can easily be extended to look for related compounds. For a collection of structures, relationships and trends can be investigated. Geometry can be obtained for a specific molecular fragment (either from a single structure or by averaging results from many structures), for use in theoretical calculations, molecular modelling, or deriving a model for Patterson search or rigid-body refinement. Trends and patterns in structures can be investigated, for example in the conformation of rings, hydrogen bonding, the structural effects of different substituents, intermolecular interactions, etc., and these can be major research investigations in their own right.

The structural databases offer a convenient route into a wealth of published literature, especially if they are used in conjunction with bibliographic databases. Their availability and reliability have made a major impact on crystallography research over the years, and they are firmly established as essential tools in the subject.

18

Other topics

18.1 Twinning

This topic was mentioned briefly in the chapter on practical aspects of structure refinement. Awareness of its existence and the problems arising from it has probably increased substantially as area-detector diffractometers have become more widespread; with serial diffractometers, twinned crystals often led to no clear and reliable unit cell and so a full set of intensity data was not measured in such cases. Data can always be collected with an area detector, even if the unit cell has not been determined, resulting sometimes in a twinned set of data.

'Twinning' is often used loosely as a description of all kinds of manifestations of samples which are not single crystals, including crystals which are slightly flawed, giving split diffraction peaks, or polycrystalline aggregates. It does, however, have a very definite meaning. A twinned crystal is one in which two (or more) orientations or mirror images of the same structure occur together in a well-defined relationship to each other. It tends to occur when there are fortuitous rational relationships among the unit cell parameters, such as for a monoclinic structure with β close to 90°, or with similar values for a and c; these are simple cases, but the possibility of twinning is by no means always obvious by visual inspection of the cell parameters, as it may involve cell diagonals or other vectors and not just the edges themselves. Twinning then results from 'mistakes' in putting composite blocks of unit cells together to form the complete crystal; some regions of the structure have a different (but related) orientation from others. Characterization of a twin requires a description of the mathematical relationship between the two unit cell orientations (called the *twin law*), and a knowledge of the relative amounts of the two or more components present. The twin law can be expressed as a 3×3 matrix that relates the two sets of unit cell axis vectors and hence also the two sets of reflection indices. The major task in sorting out a twinning problem is recognizing that it exists and finding out the twin law.

In some forms of twinning, the two component lattices coincide exactly or almost exactly, so reflections belonging to the two components are superimposed, and each measured intensity is the sum of two separate reflection intensities with different indices related by the twin law. If the twin operation is a valid symmetry operation of the crystal system but not of the point group of the crystal, this is called *merohedral twinning*. Examples of this occur in the higher symmetry crystal systems: tetragonal, trigonal, hexagonal and cubic, for each of which there are two Laue groups. The unit cell shape has the symmetry of the higher Laue group in each case. For example, a tetragonal crystal structure may belong to one of the space groups of Laue group $4/m$, in which there is no rotation or reflection symmetry associated with the a and b axes or the diagonals between them; twinning can occur by a rotation or reflection belonging to the higher Laue group 4/*mmm*, resulting in a proportion of the structure having a different orientation. This twinning operation is a symmetry operation of 4/*mmm*, but not of 4/*m*. With a specific example of reflection exchanging the *a* and *b* axes (the third *m* of 4/*mmm*), the twinning will lead to superposition of each reflection *hkl* with *khl*; these are symmetry-equivalent in the higher Laue group, but not in 4/*m*.

A particular case of merohedral twinning that can occur in any crystal system is **racemic twinning**, in which a non-centrosymmetric structure is twinned with its inverted version (all unit cells are centrosymmetric in shape, even when their detailed contents are not). This superimposes reflections *hkl* and $\overline{hkl}$, which are not strictly equivalent for a non-centrosymmetric structure when Friedel's law does not apply. This special case is considered further under the heading of anomalous dispersion (see below).

A situation similar to strictly merohedral twinning can arise in some cases when there are fortuitous rational relationships among unit cell parameters not required by the space group symmetry. It may be possible to rotate or reflect regions of the structure so that sections of the crystal lattice essentially overlap in different orientations. This is called **pseudo-merohedral twinning**, and here the twin operation is not a valid symmetry operation of the true crystal system; it is, however, a symmetry operation of a higher crystal system to which the lattice approximates. A simple example is a monoclinic structure with β close to 90°. Twofold rotation about either the *a* or the *c* axis is not valid in the monoclinic system (the only twofold rotation allowed is about *b* in the usual convention), but it is for the orthorhombic system, which has all three cell angles exactly equal to 90°; such a rotation in this case gives a unit cell of almost the same appearance and it may be possible (depending on the strength of intermolecular interactions) to build up a crystal with blocks of unit cells in the two different orientations related to each other by this rotation as the twin operation. Under monoclinic symmetry, the reflections *hkl*, $h\overline{k}l$, $\overline{h}k\overline{l}$ and $\overline{hkl}$ are equivalent (assuming Friedel's law), but the set of reflections $\overline{h}kl$, $\overline{hk}l$, $h\overline{kl}$ and $hk\overline{l}$, all equivalent to each other, are distinct from the first set. In the orthorhombic system, all eight reflections are equivalent, and this case of twinning causes the two sets to be superimposed. More complicated and less easily recognized relationships can arise and lead to pseudo-merohedral twinning.

The diffraction pattern for a merohedral or pseudo-merohedral twin can usually be indexed on a single unit cell without problems. Indications of the twinning come from problems with space group determination (such as strange or impossible systematic absences, difficulty in choosing between alternative Laue groups, or odd intensity statistics), structure solution, or refinement. Sometimes, depending on the closeness of the pseudo-merohedry, a slight splitting of reflections may be discernible, indicating that these are in fact two reflections not quite in the same place, each coming from a different component of the twin, but only an overall combined intensity can be measured for each overlapped pair (or more!) of reflections.

In other cases, the two (or more) lattices for twin components do not coincide at all points, and so only some of the reflections are superimposed or overlap. This **non-merohedral twinning** can be recognized by problems in unit cell determination, often resulting in one or more very long cell axes and many apparent absences, as an incorrectly large unit cell is generated in an attempt to fit a single reciprocal lattice to all the observed reflections, or in no cell at all unless significant proportions of the reflections are left out of the calculations. The true unit cell may be obtained either by visual inspection of the intensity-weighted reciprocal lattice (a three-dimensional display of the recorded reflections showing their positions and intensities), to identify two interpenetrating equivalent patterns in different orientations, or by use of sophisticated computer programs. The objective is to find the same unit cell in two different orientations, related to each other by a simple rotation, such that some reflections are consistent with one cell and others are consistent with the other; some reflections will fit both cells, but there should not be significant numbers of reflections that do not fit either of them. In more complex cases, there may be more than two possible orientations.

In all cases, provided some kind of starting model structure can be obtained (usually by standard Patterson or direct methods using one component of non-merohodrally twinned data or using merohedrally twinned data as if there were only one component present) and the twin law can be recognized, refinement is feasible. Each observed reflection intensity is the sum of two (or more) components and is fitted to $(F_c^2)_{\text{twin}} = k_1(F_c^2)_1 + k_2(F_c^2)_2$ (for the case of a two-part twin), where $k_2 = 1 - k_1$. Thus the twin component fraction k_1 is a refined parameter and the twin law dictates which pairs of reflections must be combined together. The detailed procedures depend on the software system used, and are different for merohedral and non-merohedral twins.

Detailed discussion of twinning, with practical illustrations, can be found in references [1–3].

18.2 Anomalous dispersion

According to Friedel's law, every diffraction pattern is centrosymmetric, even if the crystal structure is not. The structure factors $F(hkl)$ and $F(\overline{hkl})$ are complex conjugates of each other; they have equal amplitude and opposite phase, $|F(hkl)| = |F(\overline{hkl})|$ and $\phi[F(hkl)] = -\phi[F(\overline{hkl})]$. These pairs of reflections with opposite indices are known as Friedel pairs. This relationship for Friedel pairs is always true for centrosymmetric structures, but it is true for non-centrosymmetric structures only in the absence of *anomalous dispersion* (also known as *anomalous scattering*). When the X-ray wavelength is close to an *absorption edge* of a particular atom (i.e. the X-ray photon energy is similar to a difference in energy levels for electrons in the atom, so that X-ray absorption can lead to electronic excitation), this atom introduces a phase shift when it scatters the X-rays, relative to atoms of other elements. This phase shift is usually expressed by making the atomic scattering factor a complex number instead of a real number; there is a 90° out-of-phase contribution to the scattering, which may amount to the equivalent of several electrons in severe cases, and a concomitant reduction in the in-phase scattering power. The correction terms $\Delta f' + i\Delta f''$ (a real and an imaginary component for the in-phase and 90° out-of-phase components of anomalous dispersion, respectively) are wavelength-dependent and are well known for the commonly used X-ray wavelengths. Inclusion of these terms leads to a breakdown of Friedel's law for non-centrosymmetric structures containing atoms with significant anomalous dispersion effects; it is no longer true that the intensities of Friedel opposites are equal. The differences are, however, usually quite small, except close to absorption edges. Significant effects are found for atoms no lighter than silicon with Mo Kα radiation, and for atoms no lighter than oxygen with Cu Kα radiation. It is, of course, an advantage to measure Friedel opposites in the data set to optimize the available information.

Anomalous dispersion can be used to find information on reflection phases, and this is quite a commonly used method in protein crystallography, particularly if S atoms are substituted by Se, which gives a large anomalous dispersion effect with both Cu Kα and Mo Kα radiation. It is not widely used in chemical crystallography, for which standard direct methods of phase estimation normally work well (their reliability decreases as the structure size increases).

The most common use of anomalous dispersion in chemical crystallography is for the determination of *absolute configuration* of chiral molecules and for the correct resolution of related phenomena. Any non-centrosymmetric crystal structure is different from its inverse, obtained by changing the signs of all the atomic coordinates (in a few space groups, the standard choice of unit cell origin means that inversion has to be somewhere other than in the origin; see below for further details). If there is no significant anomalous dispersion, the structure and its inverse have identical diffraction patterns. In the presence of anomalous dispersion, however, the two diffraction patterns are not the same, and so it is, in principle, possible to choose the correct form of the structure corresponding to the observed diffraction pattern. In practice, the small differences can easily be swamped by experimental errors and uncertainties unless measurements are carefully made and proper corrections made for systematic effects like absorption.

Non-centrosymmetric space groups fall into a number of different types. Those with no mirror planes or glide planes (e.g. $P2_1$ or $P2_12_12_1$) can accommodate chiral molecules. In this case, distinguishing the correct structure from its inverse amounts to determining the absolute configuration of the molecule, and this is often of interest and importance, for example with natural products or in stereospecific synthesis. In polar space groups with mirror or glide planes (e.g. $Pna2_1$ or Cc), molecules of both opposite chiralities are present because of the reflection symmetry elements. What is being determined here is the direction of the polar axis, so that the crystal structure can be related to some external physical property. Finally, in space groups in some higher symmetry crystal systems (e.g. $P\bar{4}$), the experiment distinguishes different possible relative orientations of the x and y axes, a rather more subtle question that is generally not of great practical importance. In order to accommodate these different detailed phenomena, the term *absolute structure* has been suggested and is widely used, though it does not meet with universal approval.

Even if determination of the absolute configuration or other related feature is not one of the aims of a structure analysis, it is important to establish the correct absolute structure of every non-centrosymmetric crystal structure, as far as it is possible from the degree of anomalous dispersion present, because a wrong assignment affects the fit of the structural model to the observed data and, in some cases (particularly with significant anomalous scatterers present in a polar space group), leads to systematic errors in atomic positions; this is known as a polar dispersion error. In some cases, of course, the absolute configuration of a chiral molecule may be known already, for example from the synthetic route, and this can be checked against the result from the structure refinement.

There are a number of ways of deciding between the two possible forms of a non-centrosymmetric structure on the basis of anomalous dispersion effects. One method is to identify the reflections that are most strongly affected, i.e. those which should have the largest intensity differences between Friedel pairs. These can be examined to find which structure gives better agreement, and their intensities can even be remeasured with particular care to improve confidence.

Second, and more simply, the structure and its inverse can be separately refined against the observed data to see if one gives a significantly better fit in terms of R factors, standard uncertainties, etc.

A different approach is to include in the refinement a parameter that is sensitive to the effects of anomalous dispersion and can be used to discriminate between the two possible structures. One such parameter, proposed by Rogers, multiplies the imaginary part of the anomalous dispersion correction for all atoms; the correction is taken as $\Delta f' + i\eta\Delta f''$ and the parameter η is a refined parameter. A value of $+1$ with a small standard uncertainty indicates that the current structural model is correct, while -1 means the structure should be inverted. A large uncertainty shows that the absolute structure cannot be reliably determined, because the anomalous dispersion effects are too small. The more recently introduced and now much more widely used Flack parameter x is defined in terms of racemic twinning. The structure is refined as a combination of the current structural model and its inverse in proportions $1 - x$ and x respectively; x is thus a twin fraction and the twin law is the simple inversion matrix with rows $-1\,0\,0$, $0 - 1\,0, 0\,0 - 1$ as discussed in the section on twinning earlier. A correct structure gives x close to zero with a small s.u., the inverse structure gives x close to 1 with a small s.u., a large s.u. means an indeterminate result, and an intermediate value of x with a small s.u. relative to the value of x itself means the structure is probably racemically twinned. Both x and its s.u. are important in making the decision.

If it is found that the structure should be inverted, then a final refinement should be carried out with the correctly inverted model. In most cases the signs of all atomic coordinates can simply be changed (anisotropic displacement parameters remain the same on inversion), though it is common to add 1 to them all afterwards to bring the structural unit back into the 'home' unit cell; this is the same as inverting in the point $\frac{1}{2}$, $\frac{1}{2}, \frac{1}{2}$. If the space group is one of the 11 enantiomorphous pairs (e.g. $P3_1$ and $P3_2$), then the space group symmetry operators also need to be changed at the same time, because the choice of space group is itself part of the choice of handedness for the structure. There are seven rather uncommon space groups for which inversion in 0, 0, 0 or $\frac{1}{2}, \frac{1}{2}, \frac{1}{2}$ does not generate the inverted structure, because of the combination of particular symmetry elements present. The correct points of inversion are given below; $\frac{1}{2}$ may be added to any of the three numbers in each case.

$Fdd2$	$\frac{1}{8}\,\frac{1}{8}\,0$	$I4_1cd$	$0\,\frac{1}{4}\,0$
$I4_1$	$0\,\frac{1}{4}\,0$	$I\bar{4}2d$	$0\,\frac{1}{4}\,\frac{1}{8}$
$I4_122$	$0\,\frac{1}{4}\,\frac{1}{8}$	$F4_132$	$\frac{1}{8}\,\frac{1}{8}\,\frac{1}{8}$
$I4_1md$	$0\,\frac{1}{4}\,0$		

Details of these procedures and related topics can be found in references [4–8].

18.3   Sources of X-rays

Most laboratory X-ray diffractometers, whether fitted with serial detectors or area detectors, operate with a conventional sealed X-ray tube, the basic design of which has not changed for a long time, though ceramic insulating materials are increasingly being used instead of glass. The principle of operation of an X-ray tube is simple (Fig. 18.1). Electrons are generated in a vacuum by passing an electric current through a wire filament and are accelerated to a high velocity by an electric potential of tens of thousands of volts across a space of a few millimetres. The filament is held at a large negative potential and the electrons are attracted to an earthed and water-cooled metal block, where they are brought to an abrupt halt. Most of the kinetic energy of the electrons is converted to heat and carried away in the cooling water, but a small proportion generates X-rays by interaction with the atoms in the metal target. Some of the interactions produce a broad range of X-ray wavelengths, with a minimum wavelength (maximum photon energy) set by the kinetic energy of the electrons. For our purposes, however, the most important process is the ionization of an electron from a core orbital, followed by relaxation of an electron from a higher orbital to fill the vacancy. This electron transition leads to loss of excess energy by radiation, and the emitted radiation, with a definite wavelength, is in the X-ray region of the spectrum. Several different transitions are possible, so a number of intense sharp maxima (in wavelength terms) are superimposed on the broad output of the overall spectrum of radiation produced (Fig. 18.2).

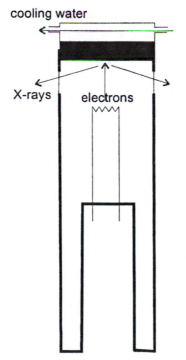

Fig. 18.1.   A schematic representation of an X-ray tube.

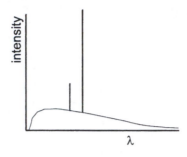

Fig. 18.2. The spectrum of X-rays emitted by an X-ray tube.

Since we use monochromatic X-rays for most purposes, one particular intense line of the output spectrum is selected and the rest discarded. The usual way of achieving this is to use diffraction itself: the X-rays emerging from a thin window in the X-ray tube are passed through a single crystal of a strongly diffracting material set at an appropriate angle. The (002) reflection of graphite is very widely used, with a 2θ angle of just over $12°$ for Mo Kα radiation ($\lambda = 0.71073$ Å, the most intense line from a target consisting of molybdenum metal) and about $26.5°$ for Cu Kα radiation ($\lambda = 1.54184$ Å, from a copper target). All other wavelengths pass undeflected through the **monochromator** crystal, leaving a single wavelength for the diffraction experiment.

Various developments of the basic X-ray tube produce higher intensities. The main limitation is the amount of heat produced, which can damage or even melt the target if it is excessive. One way to reduce the heat loading, and hence allow larger electron beam currents and more intense X-rays, is to keep the target moving in its own plane by rotating it, so that the target spot is constantly replaced. **Rotating anode** sources require continuous evacuation because of the moving parts, and their use involves much more maintenance as well as higher energy consumption than sealed tubes. The increase in intensity is usually less than a factor of ten.

Another approach is to collect and concentrate more of the X-rays generated instead of just the narrow beam taken from a standard tube. Recent technological advances have produced extremely well polished mirrors giving glancing-angle total reflection of X-rays, and devices consisting of variable-thickness layers of materials with different crystal lattice spacings to give a focusing effect through diffraction (also referred to, not strictly correctly, as mirrors). X-rays can also be concentrated and focused through glass capillaries. Each of these devices can give about an order of magnitude increase in intensity from a suitable X-ray tube (whether sealed or rotating anode). Some of them can be combined with X-ray tubes in which the electron beam is magnetically focused to give a very small target spot, reducing heat loading, and these are known as **micro-focus** tubes.

For an even greater increase in intensity, we turn to **synchrotron radiation** sources. These are national and international facilities of large size and very high cost, both in construction and in operation, and they serve many scientific, technological and commercial purposes. Examples of synchrotron sources are the SRS at Daresbury in the UK, ESRF in Grenoble (France), APS near Chicago (USA), and Spring-8 in Japan.

High intensity is not the only interesting property of synchrotron radiation (several orders of magnitude greater than from a laboratory X-ray tube). In addition, the beam is very highly collimated, it is almost completely polarized in the horizontal plane, and there is a full range of wavelengths present in a continuous spectrum, right from the infrared to hard X-rays, the exact spectral distribution depending on the operating conditions of the synchrotron.

Essentially, a synchrotron consists of a large-diameter (many metres) ring of relativistic fast-moving electrons (or positrons), which are constrained by magnetic fields to follow their circular path. This form of motion causes them to give out continuous electromagnetic radiation, the emitted energy being restored at radio frequency. In practice, very complex and expensive operating mechanisms are required for reliability, stability and safety. For monochromatic X-ray diffraction by single crystals, one wavelength is selected by a monochromator, which may also provide some focusing and compression of the beam through its special shape and construction, and further focusing and rejection of wavelength harmonics is achieved by glancing-angle reflection from long and slightly curved mirrors.

The special properties of synchrotron radiation in the X-ray region can be exploited in crystal structure analysis in a variety of ways [9].

Small crystals. Here the high incident X-ray flux is the main factor, but the possibility of focusing to produce a very fine beam is also important, the brilliance being more important than the total flux when the sample is very small. Tiny crystals occur in a very wide range of materials. A particular problem frequently encountered is of crystals which grow to a reasonable size in one or two dimensions, but form only very thin needles or plates with a total volume, and hence scattering power, below acceptable levels for laboratory study. In addition, there are applications where it is an advantage to select as small a crystal as possible to avoid other problems (e.g. in high-pressure studies, or to reduce absorption and extinction effects). The use of synchrotron radiation opens up the possibility of determining a complete structure from a single powder grain and thus investigating the homogeneity of a microcrystalline sample.

Weak scatterers. Some materials form only very small crystals because of poor crystallinity, but even large crystals may be of inferior quality, with a large mosaic spread, and so give broad and weak reflections. Synchrotron radiation can give adequate diffracted intensities, and the low intrinsic beam divergence also minimizes the breadth of observed reflections. Weak diffraction may be caused particularly by various types of disorder in the structure, and cooling of the sample is also important. A somewhat related topic is the study of substructure/superstructure relationships. Resolution of such structures depends critically on the measurement of very weak reflections which alone distinguish different possible space groups.

Unstable materials and time-resolved studies. Very high intensity means that data can be collected at maximum speed while still achieving an acceptable precision of measurement. The combination of synchrotron radiation and high-speed area detectors provides a means of doing this type of experiment. Rapid data collection is essential for unstable samples, but can also be very useful in collecting multiple data sets for a sample under

different conditions of temperature or pressure in order to investigate the effects of varying these conditions. It may be possible to follow solid-state reactions, where the reactant
and product have related structures and crystal integrity is maintained in the reaction;
these may include phase transitions, polymerization and reactions related to catalysis.

Selection of wavelength. The selection of a very short wavelength generally reduces
systematic effects such as absorption and extinction, improving the precision and accuracy of structure determination. It is also very useful in conjunction with the smaller
area detectors, because more of the diffraction pattern is available on each exposure.
Anomalous dispersion effects can be avoided or, alternatively, can be deliberately maximized and exploited by suitable wavelength selection, providing more reliable space
group discrimination, absolute structure determination for non-centrosymmetric structures, resolution of pseudo-symmetry problems, and phasing information for structure
solution. Wavelength selection is also important in high-pressure and other controlled-
environment experiments, in order to obtain the maximum amount of data when the
geometry is restricted by environment cells.

Electron density studies. A number of the above factors combine to make synchrotron
radiation particularly valuable for this type of work. Small crystals can be chosen to
minimize the effects of systematic errors such as absorption and extinction. Very short
wavelengths also help in the same way and give access to more of reciprocal space for
high-resolution data. The time taken for a data set is likely to be measured in hours rather
than the weeks required with a laboratory four-circle machine.

References

[1] R. Herbst-Irmer and G. M. Sheldrick, *Acta Cryst.*, 1998, **B54**, 443.
[2] V. Kahlenberg, *Acta Cryst.*, 1999, **B55**, 745.
[3] M. Nespolo and G. Ferraris, *Z. Krist.*, 2000, **215**, 77.
[4] H. D. Flack and G. Bernardinelli, *J. Appl. Cryst.*, 2000, **4**, 1143.
[5] H. D. Flack and G. Bernardinelli, *Acta Cryst.*, 1999, **A55**, 908.
[6] H. D. Flack, *Acta Cryst.*, 1983, **A39**, 876.
[7] P. G. Jones, *Acta Cryst.*, 1984, **A40**, 660 and 663.
[8] D. Rogers, *Acta Cryst.*, 1981, **A37**, 734.
[9] W. Clegg, *J. Chem. Soc., Dalton Trans.*, 2000, 3223.

Appendix 1

Useful mathematics and formulae

To paraphrase Lord Kelvin, when you cannot express your observations in numbers, your knowledge is of a meagre and unsatisfactory kind. The use of scientific observation to add to our knowledge inevitably means we need to express both observations and deductions mathematically. The link between the two is also mathematical. We present here some mathematics and a few formulae that are important in X-ray crystallography.

A1.1 Trigonometry

Trigonometry means 'measurement of triangles', but its use goes far beyond what its name suggests. Many properties of triangles can be summarized in terms of the ratios of the sides of the right-angled triangle in Fig. A1.1, giving:

$$\cos\theta = \frac{a}{c} \qquad \sin\theta = \frac{b}{c} \qquad \tan\theta = \frac{b}{a} \quad \text{so that } \tan\theta = \frac{\sin\theta}{\cos\theta}. \tag{A1.1}$$

The symmetry of the sine and cosine functions shows that $\cos(-\theta) = \cos(\theta)$ and $\sin(-\theta) = -\sin(\theta)$.

Another relationship among these functions is obtained from Pythagoras's theorem:

$$a^2 + b^2 = c^2 \quad \text{giving} \quad \cos^2\theta + \sin^2\theta = 1. \tag{A1.2}$$

Also useful in crystallography are the multiple angle formulae, which are given without derivation as:

$$\cos(\theta + \phi) = \cos\theta\cos\phi - \sin\theta\sin\phi$$

$$\text{and} \quad \sin(\theta + \phi) = \sin\theta\cos\phi + \cos\theta\sin\phi. \tag{A1.3}$$

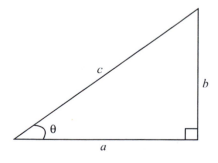

Fig. A1.1. A right-angled triangle for defining trigonometric ratios.

These come into their own in the manipulation of the electron density equation for numerical calculation. For example, by putting $\theta = 2\pi(hx + ky)$ and $\phi = 2\pi lz$ in the above expressions, $\cos 2\pi(hx + ky + lz)$ can be changed into:

$$\cos 2\pi(hx + ky + lz) = \cos 2\pi(hx + ky)\cos 2\pi lz$$
$$- \sin 2\pi(hx + ky)\sin 2\pi lz. \qquad (A1.4)$$

A similar operation gives

$$\cos 2\pi(hx + ky) = \cos(2\pi hx)\cos(2\pi ky) - \sin(2\pi hx)\sin(2\pi ky)$$
$$(A1.5)$$
$$\text{and} \quad \sin 2\pi(hx + ky) = \sin(2\pi hx)\cos(2\pi ky) + \cos(2\pi hx)\sin(2\pi ky)$$

so that

$$\cos 2\pi(hx + ky + lz)$$
$$= \cos(2\pi hx)\cos(2\pi ky)\cos(2\pi lz) - \sin(2\pi hx)\sin(2\pi ky)\cos(2\pi lz)$$
$$- \sin(2\pi hx)\cos(2\pi ky)\sin(2\pi lz) - \cos(2\pi hx)\sin(2\pi ky)\sin(2\pi lz). \quad (A1.6)$$

It looks as if we have made things far more complicated by doing this. However, these expressions usually simplify enormously in different ways according to space group symmetry and are useful in the Beevers–Lipson factorization of the electron density equation, which is how many computer programs handle Fourier transform summations.

A1.2   Complex numbers

Much of the mathematics dealing with structure factors and discrete Fourier transforms makes use of complex numbers. It is a pity these numbers have the name they do, because it has the connotation of being complicated. Complex numbers are simply numbers with two components instead of the usual one. The components are called the real and imaginary parts of the number and can be plotted on a two-dimensional diagram, called an Argand diagram, as shown in Fig. A1.2. The number plotted has real and imaginary parts of a and b respectively and can be written algebraically as $a + ib$ where $i^2 = -1$.

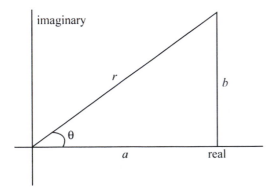

Fig. A1.2.   The complex number $a + ib$ plotted on an Argand diagram.

You may regard the imaginary constant i as a mathematical curiosity, but the important property of its square given in the previous sentence enables complex numbers to be multiplied and divided in a completely consistent way.

An equivalent way to represent a complex number is in polar form, i.e. in terms of r and θ in Fig. A1.2. These are called the modulus and argument of the number respectively. A knowledge of trigonometry allows us to write

$$a + ib = r \cos \theta + ir \sin \theta = r(\cos \theta + i \sin \theta) = re^{i\theta}. \tag{A1.7}$$

The last relationship used in equation (A1.7) is

$$\cos \theta + i \sin \theta = e^{i\theta} \tag{A1.8}$$

which is one of the most amazing relationships in the whole of mathematics. Pythagoras's theorem tells us that $r^2 = a^2 + b^2$ and we also have $\tan \theta = b/a$.

Some properties of complex numbers are important for the manipulation of structure factors. A simple operation is to take the complex conjugate, which means changing the sign of the imaginary part. Thus, the complex conjugate of the complex number $a + ib$ is written as $(a + ib)^*$ and it is equal to $a - ib$. You should be able to confirm that multiplying a complex number by its complex conjugate gives a real number which is the square of the modulus:

$$(a + ib)(a + ib)^* = (a + ib)(a - ib) = a^2 - iab + iab - i^2 b^2$$

$$= a^2 + b^2 = r^2. \tag{A1.9}$$

A1.3   Waves and structure factors

X-rays are waves and we must be able to deal with them mathematically. The obvious wavy functions are sines and cosines, so these are used in the mathematical description of waves. It is an enormous convenience to combine both sines and cosines into the single term $\exp(i\theta)$ or $e^{i\theta}$ as seen in equation (A1.8). This is the main reason why you find complex exponentials in the structure factor and electron density equations (equations (1.1) and (1.2) respectively in Chapter 1).

Similarly, the structure factors, $F(hkl)$, are the mathematical representation of diffracted waves. When they are combined to form an image of the electron density (which represents adding waves together), their relative phases are important. The mathematical construction in Fig. A1.2 allows both the amplitude of the wave, $|F(hkl)|$, and its relative phase, $\phi(hkl)$, to be represented by the modulus and argument of a single complex number. This leads us to write a structure factor in various ways such as:

$$F(\mathbf{h}) = A(\mathbf{h}) + iB(\mathbf{h}) = |F(\mathbf{h})| \cos(\phi(\mathbf{h})) + i|F(\mathbf{h})| \sin(\phi(\mathbf{h}))$$

$$= |F(\mathbf{h})| \exp(i\phi(\mathbf{h})) \tag{A1.10}$$

where the diffraction indices $(h\,k\,l)$ are represented by the components of the vector $\mathbf{h}$.

The structure factor equation, equation (1.1) in Chapter 1, shows that $F(\mathbf{h}) = F^*(\bar{\mathbf{h}})$, i.e. structure factors which are Friedel opposites are complex conjugates of each other.

This leads immediately to the relationship $F(\mathbf{h}) \times F(\bar{\mathbf{h}}) = |F(\mathbf{h})|^2$. In addition, we find that the product of any two structure factors can be written as:

$$F(\mathbf{h}) \times F(\mathbf{k}) = |F(\mathbf{h})|\,e^{i\phi(\mathbf{h})} \times |F(\mathbf{k})|\,e^{i\phi(\mathbf{k})} = |F(\mathbf{h})F(\mathbf{k})|\,e^{i(\phi(\mathbf{h})+\phi(\mathbf{k}))} \qquad (A1.11)$$

showing that the structure factor magnitudes multiply and the phases add. This is of importance when applying direct methods of phase determination.

A1.4   Vectors

A vector is often described as a quantity which has magnitude and direction, as opposed to a scalar quantity which has only magnitude. This definition is sufficient for the present purpose and we shall see how useful the directional properties of vectors are. One of the consequences of this is that vectors can be added together as shown in Fig. A1.3. The vectors $\mathbf{x}_1$ and $\mathbf{x}_{12}$ are added together to give the resultant $\mathbf{x}_2$. This is expressed algebraically as:

$$\mathbf{x}_1 + \mathbf{x}_{12} = \mathbf{x}_2. \qquad (A1.12)$$

Note that vectors are conventionally written in bold characters, as are matrices when we come to them. If the vectors $\mathbf{x}_1$ and $\mathbf{x}_2$ give the positions of two atoms in the cell, they are known as position vectors; $\mathbf{x}_{12}$ is known as a displacement vector, giving the displacement of atom 2 relative to atom 1. A rearrangement of (A1.12) expresses the displacement vector as $\mathbf{x}_{12} = \mathbf{x}_2 - \mathbf{x}_1$ and these displacement vectors arise in the description of the Patterson function (see Chapter 8).

In the unit cell, the position vector $\mathbf{x}$ has components (x, y, z) such that

$$\mathbf{x} = \mathbf{a}x + \mathbf{b}y + \mathbf{c}z \qquad (A1.13)$$

where $\mathbf{a}$, $\mathbf{b}$ and $\mathbf{c}$ are the lattice translation vectors (the edges of the unit cell) and x, y and z are the fractional coordinates of the point. The vector displacement of atom 2 from

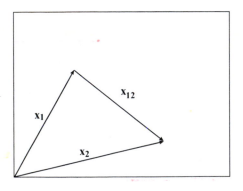

Fig. A1.3.   Addition of vectors: $\mathbf{x}_1 + \mathbf{x}_{12} = \mathbf{x}_2$.

atom 1 can therefore be written as

$$\mathbf{x}_{12} = \mathbf{x}_2 - \mathbf{x}_1 = (\mathbf{a}x_2 + \mathbf{b}y_2 + \mathbf{c}z_2) - (\mathbf{a}x_1 + \mathbf{b}y_1 + \mathbf{c}z_1)$$
$$= \mathbf{a}(x_2 - x_1) + \mathbf{b}(y_2 - y_1) + \mathbf{c}(z_2 - z_1). \tag{A1.14}$$

Similarly, the position of a point in reciprocal space is given by the vector $\mathbf{h}$, which has components (h, k, l) such that:

$$\mathbf{h} = \mathbf{a}^*h + \mathbf{b}^*k + \mathbf{c}^*l \tag{A1.15}$$

where $\mathbf{a}^*$, $\mathbf{b}^*$ and $\mathbf{c}^*$ are the reciprocal lattice translation vectors (the edges of the reciprocal unit cell) and h, k and l are usually integers giving the diffraction indices of the structure factor $F(\mathbf{h})$ at that point in the reciprocal lattice.

The scalar (dot) product of the two vectors $\mathbf{x}$ and $\mathbf{h}$ is:

$$\mathbf{h} \cdot \mathbf{x} = hx + ky + lz \tag{A1.16}$$

which is an expression to be found in both the structure factor and electron density equations. The vector (cross) product is used in the relationships between the direct and reciprocal lattices:

$$\mathbf{a}^* = \frac{\mathbf{b} \times \mathbf{c}}{V} \quad \mathbf{b}^* = \frac{\mathbf{c} \times \mathbf{a}}{V} \quad \mathbf{c}^* = \frac{\mathbf{a} \times \mathbf{b}}{V} \quad V = \mathbf{a} \cdot \mathbf{b} \times \mathbf{c} \tag{A1.17}$$

where V is the volume of the unit cell. It should be remembered that

$$\mathbf{a} \times \mathbf{b} = ab \sin \gamma \, \mathbf{n} \tag{A1.18}$$

where γ is the angle between the vectors and $\mathbf{n}$ is a unit vector perpendicular to both $\mathbf{a}$ and $\mathbf{b}$, such that $\mathbf{a}$, $\mathbf{b}$ and $\mathbf{n}$ are a right-handed set. It should be clear from (A1.17) that $\mathbf{a}^*$ is perpendicular to the bc-plane; similarly, $\mathbf{b}^*$ and $\mathbf{c}^*$ are perpendicular to the ac- and ab-planes respectively. If you need convincing that vectors are the most convenient way of expressing these relationships, here is the volume of the unit cell without using vectors:

$$V = abc(1 - \cos^2\alpha - \cos^2\beta - \cos^2\gamma + 2\cos\alpha\cos\beta\cos\gamma)^{1/2}. \tag{A1.19}$$

The angles of the reciprocal lattice can be obtained from the relationships (A1.17) but, to save you the trouble, they are:

$$\cos\alpha^* = \frac{\cos\beta\cos\gamma - \cos\alpha}{\sin\beta\sin\gamma} \tag{A1.20}$$

with corresponding expressions for $\cos\beta^*$ and $\cos\gamma^*$ obtained by cyclic permutation of α, β and γ.

A1.5   Calculation of $\sin \theta / \lambda$ for any lattice

The calculation of a Bragg angle is commonly required, for example to calculate structure factors or the setting angles on a diffractometer. In the triclinic system, the formula is:

$$\frac{4 \sin^2 \theta}{\lambda^2} = h^2 a^{*2} + k^2 b^{*2} + l^2 c^{*2} + 2hka^* b^* \cos \gamma^*$$

$$+ 2klb^* c^* \cos \alpha^* + 2lhc^* a^* \cos \beta^* \qquad (A1.21)$$

and this simplifies for other lattices.

A1.6   Determinants

Determinants feature in inequality relationships among structure factors, are needed in matrix inversion, and form a useful diagnostic tool when your least-squares refinement runs into trouble. A determinant is a square array of numbers which has a single algebraic value. An order two determinant is written and evaluated as:

$$\begin{vmatrix} a & b \\ c & d \end{vmatrix} = ad - bc \qquad (A1.22)$$

and an order three determinant is:

$$\begin{vmatrix} a & b & c \\ d & e & f \\ g & h & i \end{vmatrix} = aei + bfg + cdh - ceg - bdi - afh. \qquad (A1.23)$$

In general, a determinant can be expressed in terms of determinants of order one less than the original. For an order n determinant, this is expressed as

$$\Delta = \sum_{i=1}^{n} (-1)^{i+j} a_{ij} \Delta_{ij} \qquad (A1.24)$$

where a_{ij} is the ij element of Δ and Δ_{ij} is the determinant formed from Δ by missing out the ith row and the jth column. The summation can equally well be carried out over j instead of i and gives the same answer. However, this is only useful for determinants of small order. Evaluation of high-order determinants is best done using the process of Gauss elimination (a standard mathematical procedure not discussed here) to reduce the determinant to triangular form, then taking the product of the diagonal elements.

A1.7   Matrices

Matrices are used for a number of tasks in X-ray crystallography. Typically, they represent symmetry operations, describe the orientation of a crystal on a diffractometer, and are heavily used in the least-squares refinement of crystal structures. A brief refresher course will therefore not be out of place. A matrix is a rectangular array of numbers or algebraic expressions and matrix algebra gives a very powerful way of manipulating them.

One of the operations often required is to transpose a matrix. This exchanges columns with rows so that if $\mathbf{A}$ is the matrix

$$\begin{pmatrix} a & b \\ c & d \\ e & f \end{pmatrix},$$

its transpose, $\mathbf{A}^{\mathrm{T}}$, is

$$\begin{pmatrix} a & c & e \\ b & d & f \end{pmatrix}.$$

If a square matrix is symmetric, it is equal to its own transpose.

Matrix multiplication is carried out by multiplying the elements in a row of the first matrix by the elements in a column of the second and adding the products. This forms the element in the product matrix on the same row and column as those used in its calculation:

$$\begin{pmatrix} a & b & c \\ d & e & f \end{pmatrix} \begin{pmatrix} u & x \\ v & y \\ w & z \end{pmatrix} = \begin{pmatrix} au + bv + cw & ax + by + cz \\ du + ev + fw & dx + ey + fz \end{pmatrix}. \qquad (A1.25)$$

Multiplication can only be carried out if the number of columns in the first matrix is the same as the number of rows in the second. For example, you may wish to verify that

$$\begin{pmatrix} 2 & 3 \\ -1 & 4 \end{pmatrix} \begin{pmatrix} 3 & -2 & 1 \\ 4 & 5 & -3 \end{pmatrix} = \begin{pmatrix} 18 & 11 & -7 \\ 13 & 22 & -13 \end{pmatrix}. \qquad (A1.26)$$

Multiplication of a matrix by its own transpose always produces a symmetric matrix.

A1.8 Matrices in symmetry

Matrix multiplication is useful for representing symmetry operations. For example, the operation of the 2_1 axis relating (x, y, z) to $(\frac{1}{2} + x, \frac{1}{2} - y, -z)$ may be written as:

$$\begin{pmatrix} 1 & 0 & 0 \\ 0 & -1 & 0 \\ 0 & 0 & -1 \end{pmatrix} \begin{pmatrix} x \\ y \\ z \end{pmatrix} + \begin{pmatrix} \frac{1}{2} \\ \frac{1}{2} \\ 0 \end{pmatrix} \qquad (A1.27)$$

and this form of expression is used to represent symmetry operations in a computer.

It is sometimes useful to be able to deal with symmetry operations in reciprocal space also. The operation (A1.27) above can be written in terms of matrix algebra as

$$\mathbf{x}' = \mathbf{C}\mathbf{x} + \mathbf{d} \qquad (A1.28)$$

where $\mathbf{C}$ is the 3×3 matrix and $\mathbf{d}$ the translation vector. If a space group symmetry operation is carried out on the whole crystal, by definition the X-rays see exactly the same structure. The structure factor equation may then be written as

$$F(\mathbf{h}) = \sum_{j=1}^{N} f_j \exp(2\pi i\mathbf{h} \cdot (\mathbf{C}\mathbf{x}_j + \mathbf{d})) = \sum_{j=1}^{N} f_j \exp(2\pi i\mathbf{h}^T\mathbf{C}\mathbf{x}_j) \times \exp(2\pi i\mathbf{h} \cdot \mathbf{d})$$

$$= F(\mathbf{h}^T\mathbf{C}) \exp(2\pi i\mathbf{h} \cdot \mathbf{d}). \tag{A1.29}$$

That is, the two reflections F($\mathbf{h}$) and F($\mathbf{h}^T\mathbf{C}$) are symmetry-related. Their magnitudes are the same and there is a phase difference between them of $2\pi\mathbf{h} \cdot \mathbf{d}$.

This is easier to understand if we continue with the example in (A1.27) above. The 2_1 axis is one of those which occur in the space group $P2_12_12_1$. The symmetry-related reflections which it produces are given by

$$\mathbf{h}^T\mathbf{C} = \begin{pmatrix} h & k & l \end{pmatrix} \begin{pmatrix} 1 & 0 & 0 \\ 0 & -1 & 0 \\ 0 & 0 & -1 \end{pmatrix} = \begin{pmatrix} h & \bar{k} & \bar{l} \end{pmatrix}. \tag{A1.30}$$

That is, $F(hkl)$ is related by symmetry to $F(h\bar{k}\bar{l})$. Their magnitudes must be the same and there is a phase shift between them of:

$$2\pi\mathbf{h} \cdot \mathbf{d} = 2\pi \begin{pmatrix} h & k & l \end{pmatrix} \cdot \begin{pmatrix} \frac{1}{2} & \frac{1}{2} & 0 \end{pmatrix} = \pi(h + k). \tag{A1.31}$$

Putting this all together gives the relationships $|F(hkl)| = |F(h\bar{k}\bar{l})|$ and $\phi(hkl) = \phi(h\bar{k}\bar{l}) + \pi(h + k)$. Thus the phase is the same if $h + k$ is even, but shifted by π if $h + k$ is odd.

Even with anomalous scattering, these relationships are strictly true. It is only when structure factors are related by a complex conjugate that they are affected differently by anomalous scattering. For example, in $P2_12_12_1$, we have already seen that $|F(hkl)|$ and $|F(h\bar{k}\bar{l})|$ are always the same, but $|F(hkl)|$ and $|F(\bar{h}kl)|$ will be affected differently, as will $|F(hkl)|$ and $|F(\overline{hkl})|$.

A1.9 Matrix inversion

The inverse of the square matrix $\mathbf{A}$ is the matrix $\mathbf{A}^{-1}$ which has the property that

$$\mathbf{A}\mathbf{A}^{-1} = \mathbf{A}^{-1}\mathbf{A} = \mathbf{I} \tag{A1.32}$$

where $\mathbf{I}$ is the identity matrix (ones down the leading diagonal and zeros everywhere else).

Operations performed by multiplying by a matrix, $\mathbf{A}$, can be undone by multiplying by the inverse of the matrix, $\mathbf{A}^{-1}$. For an order 2 square matrix, the recipe for inversion is:

$$\text{if } \mathbf{A} = \begin{pmatrix} a & b \\ c & d \end{pmatrix} \quad \text{then } \mathbf{A}^{-1} = \frac{1}{\det(\mathbf{A})} \begin{pmatrix} d & -b \\ -c & a \end{pmatrix} \tag{A1.33}$$

where $\det(\mathbf{A})$ is the determinant of the matrix $\mathbf{A}$.

Inversion of an order three matrix is achieved by:

$$\text{if } \mathbf{A} = \begin{pmatrix} a_{11} & a_{12} & a_{13} \\ a_{21} & a_{22} & a_{23} \\ a_{31} & a_{32} & a_{33} \end{pmatrix} \quad \text{then form } \mathbf{C} = \begin{pmatrix} c_{11} & c_{12} & c_{13} \\ c_{21} & c_{22} & c_{23} \\ c_{31} & c_{32} & c_{33} \end{pmatrix}$$

where c_{ij} is the determinant obtained from $\mathbf{A}$ by removing the ith row and jth column and multiplying by $(-1)^{i+j}$. We then have:

$$\mathbf{A}^{-1} = \frac{1}{\det(\mathbf{A})} \mathbf{C}^{\mathrm{T}}. \tag{A1.34}$$

This recipe will work for any order of matrix, but it is extremely inefficient for orders higher than three. Larger matrices are best inverted using Gauss elimination, as mentioned earlier. It is a commonly believed fallacy that matrix inversion is necessary for solving systems of linear simultaneous equations. Since it is quicker to solve equations than to calculate an inverse matrix, the inverse should only be calculated if it is specifically required, for example to estimate standard deviations of parameters determined by the equations.

A1.10   Convolution

Convolution is an operation that affects the lives of all scientists. Since no measuring or recording instrument is perfect, it will affect the quantity that is detected before the recording takes place. For example, loudspeakers change the signal that is fed to them from an amplifier, thus altering (slightly) the sound that you hear. The mathematical description of this is called convolution. It also appears in the mathematics of crystallography, although many people function quite adequately as crystallographers without knowing much about it.

The simplest example of convolution is in the description of a crystal. The convolution of a lattice point with anything at all, e.g. a single unit cell, leaves that object unchanged. However, the convolution of two lattice points with a unit cell gives two unit cells, one at the position of each lattice point. A complete crystal, therefore, can be described as the convolution of a single unit cell with the whole crystal lattice. This would seem to be an unnecessary complication except for the intimate association of convolution with Fourier transforms.

The convolution theorem in mathematics states that: 'the Fourier transform of a product of two functions is given by the convolution of their respective Fourier transforms.' That is, if $c(x)$, $f(x)$ and $g(x)$ are Fourier transforms of $C(S)$, $F(S)$ and $G(S)$ respectively, the theorem may be expressed mathematically as:

$$\text{if } C(S) = F(S) \cdot G(S) \quad \text{then } c(x) = f(x) * g(x) \tag{A1.35}$$

where $*$ is the convolution operator.

This leads to the description of the X-ray diffraction pattern of a crystal as the product of the X-ray scattering from a single unit cell and the reciprocal lattice, seen in the pattern:

$$
\begin{array}{ccccc}
\text{unit cell} & * & \text{crystal lattice} & = & \text{crystal} \\
\updownarrow \text{F.T.} & & \updownarrow \text{F.T.} & & \updownarrow \text{F.T.} \\
\text{unit cell scattering} & \times & \text{reciprocal lattice} & = & \text{X-ray diffraction} \\
\text{pattern} & & & & \text{pattern}
\end{array}
\qquad \text{(A1.36)}
$$

This relationship allows us to deal with a single unit cell instead of the millions of cells which make up the complete crystal.

Appendix 2

A short crystallographic dictionary

We do not intend to write a complete textbook on X-ray crystallography; good ones already exist. However, it is sometimes useful to be able to look up technical terms quickly. Here are some you may come across, in alphabetical order. Terms **in bold type** in the text may be found as other entries.

Absorption

All matter absorbs X-rays. Upon passing through a slab of material without diffraction, the intensity of the X-rays varies as $I_t = I_0 \exp(-\mu t)$ where t is the thickness of material traversed and μ is the linear absorption coefficient. The value of μ depends only upon the atomic composition of the material and the X-ray wavelength. Generally, μ increases with the wavelength and the atomic number. Ignoring absorption adds systematic error to the results of a crystal structure determination. This may be significant in many cases.

Absorption edge

An absorption edge is a sudden change in the value of the linear absorption coefficient of an atom type as a function of wavelength. On the long wavelength side of the edge the absorption is low. Upon shortening the wavelength, the X-ray energy increases, allowing excitation of the atom to a higher electron energy level. This manifests itself in a sudden increase in absorption.

Anisotropy

Anisotropy is the variation of a physical parameter with direction. Thus, anisotropic displacement parameters describe the variation of atomic thermal vibrations with direction. Another example is absorption for a non-spherical crystal.

Anomalous scattering

When the X-ray frequency is close to the resonance frequency of an electron, the X-rays are scattered $\pi/2$ radians out of phase with X-rays scattered from other electrons in the same and other atoms in the crystal. This occurs when the wavelength is close to an **absorption edge** of the atom. The interference effects are such as to differentially alter the magnitudes of structure factors related as $|F(\mathbf{h})|$ and $|F(-\mathbf{h})|$, i.e. a **Friedel** pair. The size of the effect is phase dependent, so it can be used for phase determination. Its most common use in chemical crystallography is to determine the absolute configuration of a chiral molecule.

Bragg's law

Bragg's law gives the geometrical conditions under which a diffracted beam can be observed. Figure 1.7 in Chapter 1 shows rays diffracted from crystal lattice planes and, to get constructive interference, the path difference must be a whole number of wavelengths. A geometrical argument then leads to Bragg's law:

$$2d \sin \theta = n\lambda$$

where θ is known as the Bragg angle, λ is the wavelength of the X-rays, and d is the inter-plane spacing. The quantity n can always be taken as unity by an appropriate assignment of diffraction indices with a common factor of n.

Bravais lattice

The unit cell of the crystal is repeated on a regular three-dimensional space lattice to form the complete crystal. This gives one lattice point per unit cell, resulting in a so-called primitive lattice. However, it is most convenient for calculations and for the presentation of results that the symmetry of the unit cell reflects the symmetry of the crystal. This makes it necessary sometimes to have more than one lattice point per cell, resulting in centred unit cells (face centred or body centred). There are thus 14 different unit cell symmetries (Bravais lattices) which arise from all the possible different centrings in each **crystal system**.

Crystal system

There are seven crystal systems corresponding to the seven different symmetries of primitive **Bravais lattices**. Each one corresponds to different constraints on the unit cell parameters:

triclinic	no constraints
monoclinic	$\alpha = \gamma = 90°$
orthorhombic	$\alpha = \beta = \gamma = 90°$
tetragonal	$a = b, \alpha = \beta = \gamma = 90°$
rhombohedral	$a = b = c, \alpha = \beta = \gamma$
hexagonal	$a = b, \alpha = \beta = 90°, \gamma = 120°$
cubic	$a = b = c, \alpha = \beta = \gamma = 90°$

Some people will claim there are only six crystal systems because the rhombohedral system can always be represented on a centred hexagonal lattice.

Diffraction indices

Each diffracted beam (reflection) in the three-dimensional diffraction pattern is identified by three integers h, k, l, which are the diffraction indices and specify the direction in which this diffracted beam occurs. Thus, the position vector of each reflection in the **reciprocal lattice** is $h\mathbf{a}^* + k\mathbf{b}^* + l\mathbf{c}^*$. This vector is normal to the crystal planes giving rise to the diffraction and its magnitude is the reciprocal of the inter-plane spacing.

Diffraction ripple

Electron density ripple surrounding an atomic peak (particularly a heavy atom) due to limited **resolution** of the map.

Diffraction symmetry

The diffraction symmetry is the symmetry of the diffraction pattern and is described by the **Laue group**.

Discrete Fourier transform

A discrete Fourier transform is a **Fourier transform** performed on a discrete function. This is always the case when using a digital computer to perform the calculations. The X-ray diffraction pattern of a crystal is naturally discrete. The electron density, although a continuous function, is always represented on an array of points, thus making it discrete.

Enantiomorph

A crystal structure which is inverted through the origin, i.e. $\rho(\mathbf{x})$ becomes $\rho(-\mathbf{x})$, produces its enantiomorph. This does not change the magnitudes of the structure factors, but the phases are related by $\phi(\mathbf{h})$ becoming $-\phi(\mathbf{h})$. A molecule and its enantiomorph are related by a symmetry operation such as $\bar{1}$ or m.

Extinction

Two types of extinction are recognized, known as primary and secondary extinction, but their mechanisms are quite different. It is not easy to distinguish between them experimentally. Not all books give a satisfactory description, nor do they agree about which is which, so be careful!

Primary extinction

Because of faults in the crystal lattice, most crystals are thought of as consisting of a mosaic of tiny perfect crystals. These tiny crystals are all at slightly different orientations and will therefore diffract at slightly different angles. This allows the diffracted X-rays to leave the crystal without being diffracted a second time. If the angular spread of the lattice is extremely small, the probability of the X-rays being rediffracted before they leave the crystal is greatly increased. The simple (kinematic) theory of X-ray diffraction that is generally used then no longer applies and the measured intensities are considerably reduced from their expected values proportional to $|F|^2$. In an extreme case, the measured intensity can be more nearly proportional to $|F|$. The full (dynamical) theory of X-ray diffraction then needs to be used, but that is *extremely* complicated.

Secondary extinction

In a strongly diffracting crystal, the primary X-ray beam loses energy on its passage through the crystal because of the intense beams diffracted from it. The effect is an apparent increase in the absorption of the crystal by an amount proportional to the intensity of the diffracted beam. It thus has most effect on the strong reflections.

Fast Fourier transform

A computer algorithm which performs a discrete **Fourier transform** by factorizing the mathematical expression, thus increasing the efficiency of the calculation, is known as a fast Fourier transform (FFT). There are many ways of performing the factorization, but each one results in a calculating time proportional to $N \log(N)$, where N is the length of the transform. Without any factorization, the time is proportional to N^2. The first FFT algorithm was invented by Gauss.

Fourier transform

In **Fraunhofer diffraction**, the mathematical relationship between the diffraction pattern and the object which produced it is that of a Fourier transform. In the absence of extinction, the X-ray diffraction pattern is the Fourier transform of the electron density distribution in the crystal. The electron density is the inverse Fourier transform of the diffraction pattern.

Fractional coordinates

The position of a point inside the unit cell is given by the vector $\mathbf{x} = \mathbf{a}x + \mathbf{b}y + \mathbf{c}z$ where $\mathbf{a}, \mathbf{b}, \mathbf{c}$ are the lattice translation vectors and x, y, z are fractional coordinates, so-called because they can always lie in the range 0–1.

Fraunhofer diffraction

A Fraunhofer diffraction pattern is one in which all the orders of diffraction are separated. This normally occurs at an infinite distance from the diffracting object. In the case of X-ray diffraction, the orders of diffraction are separated at all distances outside the crystal because of lattice effects. The crystal X-ray diffraction pattern is thus an example of Fraunhofer diffraction.

Friedel's law

In the absence of **anomalous scattering**, the intensities in the diffraction pattern are symmetrically disposed about the origin. Specifically, the relation between $F(\mathbf{h})$ and $F(-\mathbf{h})$ is that $|F(\mathbf{h})| = |F(-\mathbf{h})|$ and $\phi(\mathbf{h}) = -\phi(-\mathbf{h})$. This is Friedel's law and the two reflections $F(\mathbf{h})$ and $F(-\mathbf{h})$ are known as Friedel opposites or a Friedel pair. Friedel's law breaks down when anomalous scattering takes place in a non-centrosymmetric crystal.

Laue group

The Laue group of a crystal is the point group symmetry of the diffracted intensities. Because of **Friedel's law**, the symmetry of the intensities is normally centrosymmetric. The **Laue group** is therefore the **point group** of the crystal with a centre of symmetry added if there was none there originally. There are 11 Laue groups corresponding to the 11 centrosymmetric crystallographic point groups.

Lorentz factor

When an X-ray intensity is measured from a moving crystal, the length of time the crystal remains in the diffracting position must be taken into account. This is done by the Lorentz factor. It depends upon the diffraction geometry and the Bragg angle.

Lp correction

The **Lorentz factor** and the **polarization factor** are normally combined into a single factor which is used to correct the measured intensities for these two effects. This is known as the Lp correction.

Metric symmetry

The symmetry of the crystal lattice is known as the metric symmetry. It may be greater than that of the space group if unit cell edges are accidentally equal or angles happen to be 90° or 120°.

Miller indices

A set of three indices, designated h, k, l, is used in classical crystallography to describe the orientation of crystal faces. They coincide with the diffraction indices when $n = 1$ in Bragg's law. Common usage blurs the distinction between the two.

Point group

A point group is a self-consistent complete set of symmetry elements which describe the symmetry about a point. The possible elements in crystallographic symmetry are the rotation axes 1, 2, 3, 4 and 6 and the inversion axes $\bar{1}, \bar{2}, \bar{3}, \bar{4}$ and $\bar{6}$. The $\bar{2}$ axis is a mirror plane and is more commonly written as m. These symmetry elements can be combined is 32 different ways, giving the 32 crystallographic point groups.

Polarization factor

X-rays produced by **Thomson scattering** are radiated anisotropically from the electron and are also polarized. This results in the diffracted X-rays being partially polarized and it also affects their intensity. The measured intensity must therefore be corrected by the polarization factor. The polarization factor depends upon the Bragg angle and also upon the source of X-rays, i.e. whether they are produced by a sealed tube or a synchrotron or have passed through a monochromator.

Reciprocal lattice

The X-ray diffraction pattern of a crystal is a discrete function that exists on a regular three-dimensional lattice, the reciprocal lattice. The reciprocal lattice can also be regarded as the **Fourier transform** of the crystal lattice. The reciprocal lattice vector $h\mathbf{a}^* + k\mathbf{b}^* + l\mathbf{c}^*$ is perpendicular to the crystal planes giving rise to the hkl reflection and has a magnitude which is reciprocal to the inter-plane spacing. The reciprocal lattice parameters can be calculated from those of the crystal lattice using the formulae given in Appendix 1.

Renninger reflection

If the crystal is in an orientation to produce two or more diffracted beams simultaneously, each diffracted beam is in the correct direction to act as the primary beam to produce a new diffracted beam. For example, if the crystal produces the hkl and $h'k'l'$ reflections at the same time, it can also give a reflection which is apparently the $h - h', k - k', l - l'$ reflection even if that is a systematic absence. There is a dynamic interchange of energy among the diffracted beams within the crystal which can lead to a grossly inaccurate measurement of the true intensity of any one of them when this effect occurs.

Resolution

Resolution is a measure of the ability to distinguish between neighbouring features in an electron density map. By convention, it is defined as the minimum plane spacing given by **Bragg's law** for a particular set of X-ray diffraction data.

Scattering factor

The atomic scattering factor gives the amount of X-ray scattering by a single atom as a function of the Bragg angle. It is known as a form factor in North America. It is evaluated as the **Fourier transform** of the electron density distribution of an atom which, in turn, is obtained from quantum mechanical calculations.

Space group

The combined configuration of a crystallographic **point group** repeated on a **Bravais lattice** is a space group. Only certain combinations of point group and Bravais lattice are self-consistent. The point group symmetry elements are generalized to include translational components in addition to rotations and inversions and they no longer must intersect at a point. A space group describes the symmetry of the atomic structure of the crystal. The total number of possible space groups is 230.

Structure factor

There are many factors affecting the intensity of X-rays in the diffraction pattern. The one which depends only upon the crystal structure is called the structure factor. It is represented by the symbol $F(\mathbf{h})$, where $\mathbf{h}$ is a **reciprocal lattice** vector, and is used to describe the X-ray diffraction pattern. It can be expressed in terms of the contents of a single unit cell according to equation (1.1) in Chapter 1.

Systematic absence

Certain classes of reflection may be absent from the diffraction pattern due to the effects of the **space group** symmetry. These are known as systematic absences and they are crucial for space group determination. They occur whenever the unit cell is centred or the symmetry elements have translational components. This allows destructive interference of the X-rays to take place in a systematic way, ensuring that certain **structure factors** must have zero magnitudes. With a centred unit cell, the systematic absences occur throughout the reciprocal lattice. Glide planes give systematic absences in the zero layer of the reciprocal lattice parallel to the glide plane. A screw axis gives systematic absences along the central reciprocal lattice row parallel to the axis.

Thermal motion

The vibration of atoms appears as a spreading out of the electron density when averaged over the whole crystal. This alters the effective X-ray scattering from each atom which may be modelled by multiplying the atomic **scattering factor** by the isotropic temperature factor $\exp(-B \sin^2 \theta / \lambda^2)$. The parameter B is related to the mean square atomic displacement U by $B = 8\pi^2 U$ and has units of Å^2. It is common to use an anisotropic model for vibrations which requires six displacement parameters instead of the single U in its description. The effect of vibrations on the diffraction pattern is to reduce its intensity at large Bragg angles. This effect is reduced when the diffraction experiment is performed at low temperature.

Thomson scattering

The mechanism by which X-rays are scattered by electrons was successfully investigated by J. J. Thomson, so the process is named after him; further description is given in Chapter 1.

Wilson plot

A plot of $\ln(\overline{|F|^2}/\sum f^2)$ versus $\sin^2\theta/\lambda^2$ which is used to determine the average value of the temperature factor and to place the observed data on an approximate absolute scale.

Appendix 3

Answers to exercises

Chapter 2

2.1 (1) Is aromatic and the most suitable solvent is likely to be aromatic also, therefore toluene (b) would be appropriate. It is also soluble in ethanol (c) and diethyl ether (g).

(2) Contains polar nitro and carboxylic acid substituents in addition to its aromatic ring. It is soluble in solvents of intermediate polarity such as ethanol (c). It is also soluble in acetic acid.

(3) Contains a large number of hydroxyl groups, making it largely insoluble in organic solvents but readily soluble in water (a).

(4) Is ionic and therefore highly soluble in water (a) and slightly soluble in ethanol (c).

(5) Contains both a long hydrocarbon chain and a carboxylic acid group. It is soluble in solvents of intermediate polarity such as dichloromethane (d), acetone (e) and diethyl ether (g).

(6) Contains aromatic substituents and toluene (b) would therefore be a good solvent.

(7) Is an organometallic compound and dichloromethane (d) is usually a good solvent for such species.

(8) Is a bicyclic saturated hydrocarbon that will possess very little polarity. The best solvent will therefore be n-hexane (f), which is similarly non-polar.

2.2 (a) This crystal is probably too small to yield useful diffraction data on a four-circle diffractometer equipped with a conventional sealed tube source. Using an area detector instrument, a more powerful source, or both, would improve the chances of success.

(b) This sample could contain crystals which belong to the cubic crystal system. Is there any supporting evidence for this, such as the crystal morphology? Such compounds are not often cubic and the possibility that you are looking at fragments of broken glass or crystals of a cubic impurity such as sodium chloride needs to be considered.

(c) This is highly suggestive of the crystal being an aggregate of smaller crystals, rendering it useless for structure determination. This can be confirmed photographically on a four-circle diffractometer or by looking at a few frames of area detector data.

(d) When tetragonal crystals are viewed along their unique c axis they do not transmit polarized light. However, they do so when viewed along any other direction. In this case the square face of each crystal is probably normal to the unique c axis.

(e) This optical behaviour suggests that the crystal has a high mosaic spread and will give reflections with wide scan widths, and diffraction data may also be of poor quality. It may be better to collect data on an area detector diffractometer.

(f) This crystal is clearly not single and you should search the sample for a better one.

Chapter 3

3.1 (a) CHBrClF: C_1 or 1 (all atoms must lie on any symmetry element).

(b) HOCl: C_s or m—note that any three-atom molecule must have a plane of symmetry!

(c) C_2H_4: D_{2h} or mmm.

(d) Z-1,2-difluoroethene: C_{2h} (2/m).

(e) E-1,2-difluoroethene: C_{2v} or $mm2$.

(f) 1R,2S-ClFHC–CHFCl: C_i ($\bar{1}$) staggered, C_s (m) eclipsed, C_1 (1) in between.

(g) 1R,2R-ClFHC–CHFCl: C_2 (2) in any plausible conformation.

(h) Tetrachlorospirane: D_2 (222): chiral and non-polar.

3.2 Symmetry elements in the crystals are as follows.

(a) No symmetry elements.

(b) Inversion centre ($\bar{1}$) at the centre of the crystal; no other symmetry elements.

(c) Twofold axis horizontally across the page.

(d) Mirror plane m normal to the page, relating two faces labelled m.

(e) Mirror plane relating faces m and c; twofold axis through face b; $\bar{1}$ at the centre of the crystal.

(f) Three twofold axes: one down the page through c, one relating two faces m, and one across the page through b.

(g) Two mirror planes: one relating faces m, and one containing the top edge of the crystal; twofold axis at the intersection of these.

(h) Three twofold axes: one down the page, one through a, and one through b; mirror plane normal to each of these; $\bar{1}$ at the centre of the crystal.

3.3 $Z = \rho N_A V/M$

(a) Methane (CH_4) [at 70 K]: $V = 215.8\,\text{Å}^3$; $\rho = 0.492\,\text{g cm}^{-3}$; $M = 16.04$; $Z = 4$.

(b) Diamond (C): $V = 45.38\,\text{Å}^3$; $\rho = 3.512\,\text{g cm}^{-3}$; $M = 12.01$; $Z = 8$.

(c) Glucose ($C_6H_{12}O_6$): $V = 764.1\,\text{Å}^3$; $\rho = 1.564\,\text{g cm}^{-3}$; $M = 180.1$; $Z = 4$

(d) Bis(dimethylgloxime)platinum(II) ($C_8H_{14}N_4O_4Pt$): $V = 1146\,\text{Å}^3$; $\rho = 2.46\,\text{g cm}^{-3}$; $M = 425.3$; $Z = 4$.

Volume (Å^3) per non-hydrogen atom: (a) $215.8/4 = 54$; (b) $45.38/8 = 5.7$; (c) $764.1/48 = 15.9$; (d) $1146/68 = 16.9$.

(a) and (b) hardly fit the definition of representative organic or organometallic molecules!

3.4 $V = 664.5\,\text{Å}^3$. For a sphere of data to a resolution r, $N = (\frac{4}{3}\pi r^{-3})/(1/V) = 5436$. Assuming that the asymmetric unit is one-eighth of this, the estimated number of data is 680. In fact, it will be more, since not all of the data (particularly $hk0$) are general.

3.5 The following tips should help you identify a cell in each case.

(a) This pattern is $p1$, with no symmetry except pure translation, despite the approximate inversion centre (as the spirals show, because they all turn in the same direction). A unit cell may be chosen of various shapes by starting on a black rectangle, selecting two neighbours and then closing up with a fourth. The asymmetric unit is the entire cell. A rectangular shape of cell can be chosen, but it is twice the size of the primitive cell, and there are no symmetry reasons for choosing it.

(b) There are mirror lines across the page (through the principal motif) and glide lines up and down the page. The plane group is thus $p2mg$. The origin is conveniently chosen on a twofold rotation point, e.g. midway between two black triangles. The unit cell is rectangular and the asymmetric unit is one quarter of it.

(c) This pattern is clearly polar, since the up direction is different from the down, and so it contains no twofold rotations. There are both mirror and glide lines up the page, and the plane group is cm. The origin is best placed on one of the mirror lines, and a rectangular cell chosen from this. The asymmetric unit is one quarter of the cell.

(d) This pattern in clearly non-polar. It contains twofold rotation points (the same as inversion centres in two dimensions), for example at the points indicated by six arrows! A rectangular unit cell may be outlined from this point. It has mirror and glide lines in both directions and is point group $c2mm$. The asymmetric unit is one-eighth of a cell.

3.6 (a) Systematic absences indicate n-glide $\perp a$ and a-glide $\perp b$. This is confirmed by the intensity distributions. The centric zone $hk0$ indicates a twofold rotation $\parallel c$. That this is a screw axis may be deduced from combining the two glide operations: $\frac{1}{2} + x, -y, z$ and $-x, \frac{1}{2} + y, \frac{1}{2} + z$ to give $\frac{1}{2} - x, \frac{1}{2} - y, \frac{1}{2} + z$ and—rather less safely—by noting that $l = 2n$ for $00l$. The space group is $Pna2_1$.

(b) Equivalence of hkl and khl data indicates that the Laue symmetry is $4/mmm$. The screw axes and the centric zones indicate $P4_12_12$ or $P4_32_12$. There would be symmetry enhancement of the rows $h00 = 0k0$ and $hh0$ by a factor of 2 and of $00l$ by a factor of four.

(c) This is the familiar puzzle: Cc or $C2/c$? There are only four molecules in the cell, which would suggest Cc, but beware!—the molecule could very well have either 2 or $\bar{1}$ symmetry and have the higher symmetry space group, which should in any case be tried first according to standard practice.

(d) This seemingly triclinic cell is betrayed by the angle very near 90°. In fact, it can be transformed by the matrix: $1\,2\,0/-1\,0\,0/0\,0\,1$ to the C-centred monoclinic cell: $a' = 12.955$, $b' = 3.952$, $c' = 9.993$ Å, $\beta = 98.42°$. This could be shown graphically by drawing the xy-plane of the old cell (which at least shows that α and γ are now 90°). The indices are converted by the same matrix, which makes the indices of the strong pairs $(5\,\bar{1}\,3)$ and $(5\,1\,3)$, $(10, 2, 0)$ and $(10, \bar{2}, 0)$, and $(\bar{4}\,\bar{4}\,0)$ and $(\bar{4}\,4\,0)$. The $0kl$ data become $2k, 0, l$ in the new cell, and the fact that these are absent for $l = 2n + 1$ shows that the space group is once again Cc or (more probably) $C2/c$.

Chapter 4

4.1 (a) $a^* = 1/a, b^* = 1/b, c^* = 1/c$, and all angles* are 90°; (b) $a^* = 1/(a \sin \beta), b^* = 1/b$, $c^* = 1/(c \sin \beta)$, $\beta^* = 180° - \beta$ ($V = abc \sin \beta$ for the monoclinic system).

4.2 $\lambda = 2d \sin \theta$; put $\sin \theta = 1$, so $\lambda = 2d$, hence $d = \lambda/2$. For Mo Kα, $d = 0.355$ Å; for Cu Kα, $d = 0.771$ Å.

4.3 $\lambda = 2 \times 0.84 \sin \theta$. For Mo K$\alpha$, $\sin \theta = 0.71073/1.68$, so $\theta = 25.0°$; for Cu Kα, $\sin \theta = 1.54184/1.68$, so $\theta = 66.6°$.

4.4 (a) For $\psi = 0$, $\phi_0 = \tan^{-1}(-x/y)$

         $\chi_0 = \tan^{-1}(z/k(x^2 + y^2))$, where $k^2 = 1/(x^2 + y^2)$,

         so $\chi_0 = \tan^{-1}(z/\sqrt{(x^2 + y^2)})$

         $\theta_0 = \sin^{-1}(\lambda r/2) = \sin^{-1}(\lambda\sqrt{(x^2 + y^2 + z^2)}/2)$

         $\omega_0 = \theta_0$

(b) For $\psi = 90°$,   the second term of ϕ has zero denominator, so $\phi_0 - 90°$

         the denominator for χ is also zero because of the 90° difference

         between ϕ and ϕ_0, so $\chi = \pm 90°$ depending on the sign of z

         $\theta = \theta_0$ (this is independent of ψ)

         ω is not so easy to simplify!

4.5 $a = 8, b = 4, c = 10$ Å. The crystal is mounted exactly about the c axis. This is not a good idea because of multiple reflection effects.

4.6 200: 14.36, 7.18, 0, 0
020: 28.96, 14.48, 0, −90
002: 11.48, 5.74, −90, 0
42$\bar{5}$: 51.32, 25.66, 35.26, −45.00

4.7 200: 11.48, 5.74, −90, −90, should be accessible
42$\bar{5}$: 51.32, −29.08, 90, −135.00, this could be blocked by the χ circle
200 and 020 are almost certainly inaccessible

4.8   Indices are 111.

4.9 $\lambda = 2d \sin \theta$; $d_{002} = c/2 = 5$ Å. For $\lambda = 1$ Å, this gives $\theta = 5.739°$. Incorrect $\theta = 5.714°$, which on reverse calculation gives $d_{002} = 5.022$ Å, so the incorrect $c = 10.044$ Å, a 0.44% error.

Chapter 5

5.1   (a) Although absorption is lower with Mo radiation, the difference in this case is rather small (about 20%). If the crystal is weakly diffracting, you should use Cu radiation in order to obtain stronger data.

      (b) Absorption is more than twice as serious with Cu. This is due to the high chlorine content: bromine has similar absorption with the two types of radiation but chlorine absorbs Cu radiation more strongly than Mo. Therefore, Mo is clearly better from this point of view.

      (c) Ru lies beyond the absorption edge for Mo and consequently the absorption has dropped off, so Mo is strongly preferred, as absorption effects will be much less serious than with Cu.

      (d) This experiment requires the use of Cu radiation, as the anomalous scattering of N and O is insignificant with Mo.

      (e) Either could be used, as the presence of bromine atoms will lead to strong anomalous effects with both types of radiation. Cu radiation may be preferable if the crystal diffracts weakly.

5.2   This question is intended to initiate discussion. For example, why is the crystal poorly diffracting? Is it very small, in which case copper radiation is a possibility? If it is a normal-sized crystal, then copper radiation is less practical. If conventional sources and detectors are inadequate, then rotating anodes, synchrotron radiation and area detectors should be considered. Do not forget that growing a better crystal is an option!

5.3   This is a highly artificial example and these symptoms would not all occur with the same crystal. Note that there is a good spread of angles in χ and φ, so that should not be a problem. The warning signs are:

      reflections 3 and 11 are very weak and should be omitted;

      reflection 4 is very strong (it can be omitted temporarily but must ultimately be indexed);

      reflection 7 is anomalously sharp (possibly an instrumental artefact or double diffraction);

      reflections 10 and 11 are very wide (noise of some kind, or is the crystal unsuitable?);

      reflections 12 and 15 have very similar setting angles; is more than one crystal present?

5.4   If this pattern is repeated throughout the full set of measurements, the answer is no. The reflections divide into two clearly distinct sets of monoclinic equivalents with intensities grouped around 260 and 200.

5.5   There is no mathematically unique answer to this question. The compound and THF have 37 and five non-H atoms requiring 666 and 90 Å^3 respectively. The unit cell could contain three molecules of the compound (1998 Å^3) and no THF, but the agreement is not good and $Z = 3$ is unlikely. If it held two molecules of the compound (1332 Å^3), that would leave 518 Å^3, enough for about six molecules of THF per unit cell. Such a crystal would be likely to suffer a rapid loss of solvent unless protected, for example by handling it under mother liquor and collecting data with the crystal coated in a film of frozen oil.

5.6   For each index h, k and l, sum the absolute values and divide by the corresponding unit cell length. Are the ratios broadly similar, say within a factor of two? If not, the choice of reflections is not suitable. Using all reflections the ratios are approximately 2, 3 and 6 so the choice is not satisfactory. You will notice that the fourth reflection $(-3, 10, 1)$

is expected to be systematically absent in a C-centred cell and is therefore definitely not suitable. Repeating the calculation with $(-3, 10, 1)$ omitted gives ratios of 2, 2 and 6 which is similarly unsatisfactory. The list of reflections needs more entries having higher indices of h and k.

5.7 (a) The first reflection has a rather high value of 2θ and could therefore be affected by $\alpha_1 - \alpha_2$ splitting (whether this will pose a problem depends on the instrument being used). The second reflection is more worrying: it is a strong, low-angle reflection which may gain intensity as irradiation of the crystal increases its mosaic spread. This behaviour is unlikely to be typical of the dataset as a whole and applying a drift correction based on the behaviour of such a reflection is likely to reduce the quality of the dataset.

(b) The second and third reflections are too close together to provide independent checks on crystal movement (the angle between their vectors is about 30°). Standard reflections should be chosen to be as orthogonal as possible to each other (i.e. the angle between their vectors should be as close as possible to 90° and certainly never less than 45°).

(c) The reflections are suitably orthogonal but the third is weak in comparison to the others and it may show large relative variations in intensity, making it unsuitable for either monitoring the behaviour of the crystal or applying drift corrections during data reduction. It is impossible to be certain unless we know its significance level (standard reflections should normally have $(I)/\sigma(I)$ ratios of at least 50).

5.8 Consider the likely paths the beams will follow and calculate $(I/I_0) = e^{-\mu x}$ for each case. The minimum and maximum path lengths are as shown in the first column below, and the corresponding values of $e^{-\mu x}$ are shown in the three following columns. For the crystal mounted with its needle axis vertical (c), it is possible to make the reasonable assumption that no X-ray beam will pass along this axis and the minimum and maximum path lengths are 0.06 and 0.08 mm, respectively.

Min and max path lengths (mm)	Range of (I/I_0) for		
	$\mu = 0.1\,\text{mm}^{-1}$	$\mu = 1.0\,\text{mm}^{-1}$	$\mu = 5.0\,\text{mm}^{-1}$
(a) 0.02, 0.40	0.998–0.961	0.980–0.670	0.904–0.135
(b) 0.20, 0.40	0.980–0.961	0.818–0.670	0.368–0.135
(c) 0.06, 0.08	0.994–0.992	0.942–0.923	0.741–0.670
(d) 0.06, 0.40	0.994–0.961	0.942–0.670	0.741–0.135

Note that for crystals with $\mu = 0.1\,\text{mm}^{-1}$ absorption is not significant and, irrespective of crystal morphology, the range of transmission coefficients is very narrow. With higher values of μ it is obvious that absorption is particularly extreme for the thin plate. Mounting an acicular crystal with its needle axis vertical (c) is better than mounting it horizontally (d) in that the former gives a much narrower range of transmission coefficients, especially with higher values of μ.

5.9 This calculation involves some approximations and assumptions, but the point is to get a reasonable estimate of the uncertainties involved and decide whether these are important. First consider case (b) and calculate the relative uncertainties in the three cell dimensions (these are approximately the same at about 1 part in 2000). Then proceed on the assumption that cell errors are isotropic. The value of 1 in 2000 gives a contribution to the uncertainty in the C–C bond of $1.520\,\text{Å}/2000 = 0.0008\,\text{Å}$, which will not be significant in any but the most accurate determinations. The calculation is valid for all orientations of the C–C bond. The cell in (a) is obviously poorer. Calculate the relative uncertainties in a, b and c: the values are 1 in 700, 1 in 650 and 1 in 350, respectively, being much higher and not isotropic. Calculate the contribution to the uncertainty for a C–C bond lying parallel to each of the [100], [010] and [001] directions. The answers are 0.002, 0.002 and 0.004 Å, respectively. These, especially the last, would add significantly to the uncertainty of a typical structure determination. The calculated values are probably underestimated, as we have ignored any contribution from the unconstrained angles.

5.10   (a) 280(33); (b) 2016(54); (c) −14(21); (d) 52(16).

Chapter 7

7.1   Let c represent $\cos 2\pi$ and s represent $\sin 2\pi$ for brevity. The cosine term (A) factorizes to $chx \cdot c(ky+lz) - shx \cdot s(ky+lz)$, and then further to $chx \cdot cky \cdot clz - chx \cdot sky \cdot slz - shx \cdot cky \cdot slz - shx \cdot sky \cdot clz$. In a similar manner, the sine term (B) may be factorized twice to give $shx \cdot cky \cdot clz + chx \cdot sky \cdot clz + chx \cdot cky \cdot slz - shx \cdot sky \cdot slz$. Since $sx = -s(-x)$ and $cx = c(-x)$, the inversion centre will cancel out all values with odd numbers of sine terms, so B = 0. The remaining parts of A which have two sine terms will also vanish; for example, x,y,z will cancel x, y, $-z$ in any term containing slz. The entire expression thus reduces to B = 0 and A = $8chx \cdot cky \cdot clz$.

7.2   As the cell edge in projection is 2.72 Å, the length of the body diagonal will be 4.71 Å (multiply by $\sqrt{3}$). The peaks corresponding to S atoms are at 0.22 and 0.78, so two S atoms lying on the same threefold axis must be 0.44 × 4.71 or 0.56 × 4.71 Å apart, i.e. 2.07 or 2.67 Å. The former is the more plausible. The Fe · · · S distance may be calculated by assuming that the distance from Fe to the midpoint of the S–S bond will be 2.72 Å (half a cell edge) along, say, the x direction. Thus, with reference to it, the coordinates of the nearest S will be 0.50–0.11, 0.11, 0.11 referred to the true unit cell with an edge of 5.43 Å, so the distance is 2.28 Å.

7.3   This problem may be best tackled by a group of people together, so that the calculations can be shared out! The 'Patterson' gives a peak at $z = 0.197$ (sum = 88 at 0.19 and 97 at 0.20). Thus, the signs for the Fourier may be taken from a Br position at $z = 0.10$, making all signs positive there. When this sum is carried out, there is a detectable pattern of regular peaks showing the tail of CH_2 groups, which overlap in pairs in this projection, so there are half the number of peaks as the number of atoms in the chain.

Chapter 8

8.1   Of the 16 interatomic vectors for four atoms, four coincide with the origin; two coincide at each of $2x, \frac{1}{2},\frac{1}{2} + 2z$ and $-2x, \frac{1}{2}, \frac{1}{2} - 2z$ (the Harker plane); two at each of $0, \frac{1}{2} + 2y, \frac{1}{2}$ and $0,\frac{1}{2} - 2y, \frac{1}{2}$ (the Harker line); and the final four occupy the general positions $2x, 2y,$ $2z; 2x, -2y, 2z; -2x, 2y, -2z; -2x, -2y, -2z$ in the Patterson space group $P2/m$.

8.2   There are 32 equally satisfactory positions for the unique Ru atom, depending on how the observed peak positions are compared with the full list from Exercise 8.1, including symmetry equivalents, and on the usual ambiguity whereby $\frac{1}{2}$ can be added to or subtracted from any coordinate. The most obvious solution is 0.224, 0.222, 0.081. If the molecule is to be a dimer, this Ru atom must be bridged by 2 Cl atoms to another Ru related to it by an inversion centre (if it is bridged somehow to one related by a screw axis or glide plane, the result is an infinite polymer). The nearest suitable symmetry-related atom to this one is the one at −0.224, −0.222, −0.081, and it will be about $[\sqrt{(0.45^2 + 0.44^2 + 0.16^2)}] \times 15$ Å or about $9.5\,\text{Å}$ away. This is much too far, as an Ru–Cl–Ru bridge will be not more than $3.5\,\text{Å}$.

8.3 The simplest method is probably to place the Ni and bridging 'S' atoms at the corners of a square, thus:

Ni1 at 1.2, 1.2, 0	S1 at 1.2, −1.2, 0
Ni2 at −1.2, −1.2, 0	S2 at −1.2, 1.2, 0

The other S atoms may then be placed at

1.2, 1.2, 1.2	1.2, 1.2, −1.2
−1.2, −1.2, 1.2	−1.2, −1.2, −1.2
1.2, 2.4, 0	2.4, 1.2, 0
−2.4, −1.2, 0	−1.2, −2.4, 0

8.4 The systematic absences show a C-centred unit cell and a c-glide perpendicular to b in every case (the third symmetry operation). $Cmc2_1$ will have a Harker plane perpendicular to c ($2x$, $2y$, $\frac{1}{2}$) and a Harker line parallel to a ($2x$, 0, 0). $C2cm$ will have a Harker plane perpendicular to a (0, $2y$, $2z$) and a Harker line parallel to c (0, 0, $\frac{1}{2} + 2z$).

Chapter 9

9.1 The required determinant is

$$\begin{vmatrix} E(0) & E(\mathbf{h}) & E(2\mathbf{h}) \\ E(-\mathbf{h}) & E(0) & E(\mathbf{h}) \\ E(-2\mathbf{h}) & E(-\mathbf{h}) & E(0) \end{vmatrix}$$

which can be expanded to give the inequality relationship

$$E(0)[E^2(0) - |E(2\mathbf{h})|^2 - 2|E(\mathbf{h})|^2] + 2|E(\mathbf{h})|^2 E(2\mathbf{h}) \geq 0.$$

This can be simplified by cancelling out a common factor of $[E(0) - E(2\mathbf{h})]$ and rearranging to give

$$|E(\mathbf{h})|^2 \leq \tfrac{1}{2} E(0)[E(0) + E(2\mathbf{h})].$$

With the given amplitudes, the left-hand side of the inequality is 4 and the right-hand side is $\frac{15}{2}$ or $\frac{3}{2}$ for $E(2\mathbf{h})$ positive or negative respectively. The sign of $E(2\mathbf{h})$ must, therefore, be positive.

9.2 Equation (9.8) comes directly from the expansion of the determinant in (9.7). With the given amplitudes, the inequality becomes $8 \geq 0$ or $-8 \geq 0$ depending on the sign of $E(-\mathbf{h})E(\mathbf{h} - \mathbf{k})E(\mathbf{k})$. The sign of $E(-\mathbf{h})E(\mathbf{h} - \mathbf{k})E(\mathbf{k})$ must, therefore, be positive.

9.3 The order four determinant is

$$\begin{vmatrix} E(0) & E(\mathbf{h}) & E(\mathbf{h}+\mathbf{k}) & E(\mathbf{h}+\mathbf{k}+\mathbf{l}) \\ E(-\mathbf{h}) & E(0) & E(\mathbf{k}) & E(\mathbf{k}+\mathbf{l}) \\ E(-\mathbf{h}-\mathbf{k}) & E(-\mathbf{k}) & E(0) & E(\mathbf{l}) \\ E(-\mathbf{h}-\mathbf{k}-\mathbf{l}) & E(-\mathbf{k}-\mathbf{l}) & E(-\mathbf{l}) & E(0) \end{vmatrix}.$$

With $E(\mathbf{h}+\mathbf{k}) = E(\mathbf{k}+\mathbf{l}) = 0$, this forms the inequality relationship

$$E^2(\mathbf{0})[E^2(\mathbf{0}) - |E(\mathbf{h})|^2 - |E(\mathbf{k})|^2 - |E(\mathbf{l})|^2 - |E(-\mathbf{h}-\mathbf{k}-\mathbf{l})^2|] + |E(\mathbf{h})E(\mathbf{l})|^2$$
$$+ |E(\mathbf{k})E(\mathbf{h}+\mathbf{k}+\mathbf{l})|^2 - 2E(\mathbf{h})E(\mathbf{k})E(\mathbf{l})E(-\mathbf{h}-\mathbf{k}-\mathbf{l}) \geq 0$$

and with suitably large amplitudes this can be used to prove that the sign of $E(\mathbf{h})E(\mathbf{k})E(\mathbf{l})E(-\mathbf{h}-\mathbf{k}-\mathbf{l})$ must be negative; this is a negative quartet relationship.

9.4　The three determinants are obtained from the one in Exercise 9.3 by cyclic permutation of indices. Together, they give a stronger indication of the negative quartet provided that $E(\mathbf{h}+\mathbf{k}) = E(\mathbf{k}+\mathbf{l}) = E(\mathbf{h}+\mathbf{l}) = 0$.

9.5

Determined signs

To get started, arbitrary signs (+) have been assigned to 5,7 and 14,5, and the symbol A to 8,8. B means the opposite sign to A. Alternative solutions may be obtained by starting with other cominations of signs. The fact that $A = +$ is shown by the alternative values found for 8.13, here B and −.

								Determined signs
								1 10 B
								1 14 BBB
5	7 +	−5	7 +	5	7 +	14	5 +	2 17 A
5	−7 +	14	5 +	10	0 +	−9	12* −	
10	0 +	9	12 +	15	7 +	5	17 −	2 19 −
5	17 −	−5	7 +	14	−5* −	5	7 +	3 10 BB
5	−17 −	8	8 A	−8	8 A	6	−3* A	
10	0 +	3	15 A	6	3 B	11	4 A	3 15 A
−5	7 +	9	−12* −	5	17 −	−5	17 −	5 7 +
6	3 B	−3	15 A	6	−3* A	6	−3* A	
1	10 B	6	3 B	11	14 B	1	14 B	5 17 −
11	14 B	−1	14* A	−1	10* A	14	5 +	5 19 B
−10	0 +	10	0 +	8	−8 A	−7	−2 +	
1	14 B	9	14 A	7	2 +	7	3 +	6 3 BB
5	7 +	−5	17 −	11	−4* B	14	5 +	7 2 ++
7	3 +	7	2 +	1	14 B	−9	14* B	
12	10 +	2	19 −	12	10 +	5	19 B	7 3 +++
5	19 B	11	−4* B	−3	15 A	9	9 A	7 10 AA
5	−19 B	1	10 B	12	−6 +	−8	8 A	
10	0 +	12	6 +	9	9 A	1	17 +	8 8 A
6	3 B	6	3 B	−5	7 +	9	−12* −	8 13 B −
6	3 B	7	3 +	13	6 B	1	17 +	
12	6 +	13	6 B	8	13 B	10	5 −	9 9 A
−3	15 A	−5	7 +	5	−7 +	−7	10* B	9 12 +
10	−5* +	7	10 A	−2	17* B	10	0 +	
7	10 A	2	17 A	3	10 B	3	10 B	9 14 A

10	5 –	9	–9 A	5	19 B	–5	19 B		10	0 +++
–9	9 A	–2	19*+	2	–17*B	13	–6*A			
1	14 B	7	10 A	7	2 +	8	13 –		10	5 –

–2	17*B	–1	–10 B		11	4 A
9	–14*B	8	13 B			
7	3 +	7	3 +		11	14 B

12 6 ++

If you had to look at the solution in order to decide what to do,
try again with another starting set, say 5, 7 = + and   12 10 ++
14, 5 = –. You should still get a consistent set, and the
symbol A should still come out to be +.   13 6 B

14 5 +

15 7 +

Chapter 11

11.1   The expected error in α is half that of the others, so the weight is twice that of the others:
instead of $\alpha = 73$ we have $2\alpha = 146$. The stronger application of the restraint changes the
equation $\alpha + \beta + \gamma = 180$ into $2\alpha + 2\beta + 2\gamma = 360$ (the factor of 2 is arbitrary).
The normal equations are $\mathbf{A}^T\mathbf{A}\mathbf{x} = \mathbf{A}^T\mathbf{b}$, i.e.

$$\begin{pmatrix} 2 & 0 & 0 & 2 \\ 0 & 1 & 0 & 2 \\ 0 & 0 & 1 & 2 \end{pmatrix}\begin{pmatrix} 2 & 0 & 0 \\ 0 & 1 & 0 \\ 0 & 0 & 1 \\ 2 & 2 & 2 \end{pmatrix}\begin{pmatrix} \alpha \\ \beta \\ \gamma \end{pmatrix} = \begin{pmatrix} 2 & 0 & 0 & 2 \\ 0 & 1 & 0 & 2 \\ 0 & 0 & 1 & 2 \end{pmatrix}\begin{pmatrix} 146 \\ 46 \\ 55 \\ 360 \end{pmatrix}$$

which gives

$$\begin{pmatrix} 8 & 4 & 4 \\ 4 & 5 & 4 \\ 4 & 4 & 5 \end{pmatrix}\begin{pmatrix} \alpha \\ \beta \\ \gamma \end{pmatrix} = \begin{pmatrix} 1012 \\ 766 \\ 775 \end{pmatrix}.$$

Confirm the solution $\alpha = 73.6°$, $\beta = 48.4°$, $\gamma = 57.4°$ by showing this satisfies the
equations.

11.2   Observational equations are

$$\begin{pmatrix} 1 & 1 \\ 3 & 1 \\ 5 & 1 \end{pmatrix}\begin{pmatrix} m \\ c \end{pmatrix} = \begin{pmatrix} 2 \\ 3 \\ 7 \end{pmatrix}.$$

Normal equations are

$$\begin{pmatrix} 1 & 3 & 5 \\ 1 & 1 & 1 \end{pmatrix}\begin{pmatrix} 1 & 1 \\ 3 & 1 \\ 5 & 1 \end{pmatrix}\begin{pmatrix} m \\ c \end{pmatrix} = \begin{pmatrix} m \\ c \end{pmatrix} = \begin{pmatrix} 1 & 3 & 5 \\ 1 & 1 & 1 \end{pmatrix}\begin{pmatrix} 2 \\ 3 \\ 7 \end{pmatrix}$$

which gives

$$\begin{pmatrix} 35 & 9 \\ 9 & 3 \end{pmatrix} \begin{pmatrix} m \\ c \end{pmatrix} = \begin{pmatrix} 46 \\ 12 \end{pmatrix}$$

and the solution is $m = \frac{5}{4}, c = \frac{1}{4}$, so the line of regression is $y = \frac{5}{4}x + \frac{1}{4}$.

11.3   The matrix of normal equations is

$$\begin{pmatrix} 35 & 9 \\ 9 & 3 \end{pmatrix}.$$

Its inverse is

$$\frac{1}{24} \begin{pmatrix} 3 & -9 \\ -9 & 35 \end{pmatrix}.$$

This gives values proportional to

$$\begin{pmatrix} \sigma_m^2 & \sigma_m \sigma_c \mu_{mc} \\ \sigma_m \sigma_c \mu_{mc} & \sigma_c^2 \end{pmatrix}$$

so that $\mu_{mc} = -9/\sqrt{(3 \times 35)} = -0.86$.

11.4   The weighted observational equations are

$$\begin{pmatrix} 2 & 0 & 0 \\ 0 & 1 & 0 \\ 0 & 0 & 2 \\ 1 & 1 & 1 \end{pmatrix} \begin{pmatrix} \alpha \\ \beta \\ \gamma \end{pmatrix} = \begin{pmatrix} 146 \\ 46 \\ 110 \\ 180 \end{pmatrix}.$$

The normal equations are

$$\begin{pmatrix} 5 & 1 & 1 \\ 1 & 2 & 1 \\ 1 & 1 & 5 \end{pmatrix} \begin{pmatrix} \alpha \\ \beta \\ \gamma \end{pmatrix} = \begin{pmatrix} 472 \\ 226 \\ 400 \end{pmatrix}.$$

11.5   All equations are linear except the cosine rule. Write this as

$$f(\alpha, \beta, \gamma, a, b, c) = b^2 + c^2 - a^2 + 2bc \cos \alpha = 0$$

then the derivatives are

$$\frac{\partial f}{\partial \alpha} = -2bc \sin \alpha \qquad \frac{\partial f}{\partial \beta} = 0 \qquad \frac{\partial f}{\partial \gamma} = 0$$

$$\frac{\partial f}{\partial a} = -2a \qquad \frac{\partial f}{\partial b} = 2b + 2c \cos \alpha \qquad \frac{\partial f}{\partial c} = 2c + 2b \cos \alpha.$$

The matrix of derivatives is therefore

$$\begin{pmatrix} 1 & 0 & 0 & 0 & 0 & 0 \\ 0 & 1 & 0 & 0 & 0 & 0 \\ 0 & 0 & 1 & 0 & 0 & 0 \\ 0 & 0 & 0 & 1 & 0 & 0 \\ 0 & 0 & 0 & 0 & 1 & 0 \\ 0 & 0 & 0 & 0 & 0 & 1 \\ -2bc \sin \alpha & 0 & 0 & -2a & 2b + 2c \cos \alpha & 2c + 2b \cos \alpha \\ 1 & 1 & 1 & 0 & 0 & 0 \end{pmatrix}.$$

Chapter 12

12.1 (a) This weights up high-angle data, which contain little contribution from H atoms or deformation density. So it is useful for obtaining atomic positions that approximate better to nuclear positions; then difference maps will show H atoms and/or non-spherical distributions more clearly.

(b) This weights up low-angle data, which depend more on gross features of the structure. It may help in the initial stages of structure development. A more drastic and more efficient method is to apply a θ_{max} cut-off.

12.2 (a) 12 atoms $\times$ (3 coordinates $+ 1 U_{iso}$) $= 12 \times 4 = 48$; plus overall scale factor gives 49.

(b) $12 \times (3 + 6) + 1 = 109$.

(c) $12 \times 9 + 12 \times 4$ (for H atoms) $+ 1 = 157$.

12.3 (a) y is free, but x and z are fixed at appropriate values. $U^{13} = U^{23} = 0$, other U^{ij} are free.

(b) z is free, but x and y are linearly related (e.g. $x = y$, but the exact constraint depends on the precise position of the mirror plane). For the position x,x,z $U^{11} = U^{22}$ and $U^{13} = U^{23}$.

12.4 (a) none; (b) all; (c) a and c; (d) b.

12.5 A molecule of THF is not planar!

12.6 (a) All C−F equal, all F $\cdots$ F equal (i.e. all F−C−F angles equal), rigid-bond U^{ij} restraints for all C−F bonds.

(b) All C−C bonds equal, all C−C−C angles equal (equivalent to making all 1,3-distances equal), all 1,4-distances equal (distinction from the boat form, where this is not true), rigid-bond U^{ij} for C−C bonds. Riding model constraints should also be used for the H atoms.

12.7 Refining two symmetry-equivalent parameters as if they were independent gives a singular matrix, which cannot be inverted (the determinant is zero). Pseudo-symmetry means the matrix determinant is small (the matrix is ill-conditioned), so inversion produces large parameter shifts and large uncertainties.

12.8 Examples are shown in Fig. A3.1.

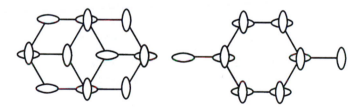

Fig. A3.1. Possible modes of disorder for toluene across an inversion centre. In each case, two overlapping positions are shown, with ellipses elongated horizontal and vertical for the component atoms.

12.9 The space group symmetry operators themselves are chiral. It is necessary to change to space group $P4_3$, which has operators of the opposite chirality.

12.10 The displacement parameter(s) become very large in an attempt to smear out the unwanted electron density. Also, R values are larger than without the atom.

12.11 (a) R indices usually decrease (more so if unweighted), but s.u.s probably increase.

(b) R indices decrease, s.u.s decrease, because of the better fit.

(c) R indices increase, so do s.u.s.

(d) R indices decrease, probably not by much; there is little effect on s.u.s.

(e) R indices decrease, so do s.u.s.

Chapter 13

13.1 $\bar{x} = 1.532, \sigma = 0.004, \sigma(\bar{x}) = 0.001$ Å.

13.2 $\sigma^2(x_1 + x_2) = \sigma^2(x_1) + \sigma^2(x_2)$

$\sigma^2(x_1 - x_2) = \sigma^2(x_1) + \sigma^2(x_2)$

$\sigma^2(x_1 + x_2 + x_3) = \sigma^2(x_1) + \sigma^2(x_2) + \sigma^2(x_3)$

$\sigma^2(x_1 x_2) = x_2^2 \sigma^2(x_1) + x_1^2 \sigma^2(x_2)$

13.3   The sum of the angles is $180°$ exactly; its uncertainty is zero—an example of total correlation!

Chapter 14

14.1   (a) no; (b) yes, very probably; (c) probably not.

14.2 $\chi^2 = 42.65$, which is too large for a set of values taken from a single population, so the bond lengths are *not* all equivalent (assuming the s.u.s are correct!).

14.3   (a) planar, 2; (b) non-planar, 2; (c) planar, 2; (d) non-planar, 2; (e) planar, 4.

14.4   Three lengths (each equal to the opposite one), and three angles (all the other 12 are either the same as these, $(180° - \text{these})$, or exactly $180°$.

14.5   Six lengths. Strictly speaking there are nine independent angles (because there are $15 = (3 \times 7) - 6$ internal degrees of freedom for 7 atoms), but all 15 should normally be quoted!

Chapter 15

15.1   (a) C(1) is vibrating rather less than its neighbours and N(3) somewhat more. It is possible that these atoms have been assigned the wrong atom types (and therefore scattering factors) and that they should be N(1) and C(3) respectively. Examination of the U^{ij} values may show the difference more clearly and the molecular geometry around each atom should be inspected. A difference electron density plot in the plane of the ring may show unmodelled electron density. Both models should be refined to discover whether one is clearly better. If neither is satisfactory, disorder may be present, involving a rotation of $180°$ about an axis passing through C(6), C(2) and the midpoint of the C(4)–C(5) bond. Disorder modelling will probably require restraints to be applied to the ring geometry: one possibility is similarity restraints so that the two ring components have the same internal geometry.

(b) It appears that C(9) is affected by disorder, which is being treated (by default) as dynamic disorder about a single atom site. This may not be the best description of the disorder and a static model involving two or more alternative sites for the atom may be more appropriate. In order to evaluate the static model it is first necessary, by analysis of its U^{ij} values, to 'split' C(9) to give two alternative positions [C(9A) and C(9B)]. Restraints are useful here, especially geometric ones which either fix both terminal C–O distances to a standard length or restrain C(9A)–O(9) and C(9B)–O(9) to be equal. It may only be possible to refine the disorder components with isotropic displacement parameters; alternatively it may be possible to refine anisotropic displacement parameters provided rigid-bond restraints are applied to these. It is not correct in general to assume that the sites of C(9A) and C(9B) are equally occupied and an occupancy parameter must be refined: with two disorder components only one parameter is needed, as the occupancies of the two sites must sum to unity. When both dynamic and static disorder models have been fully investigated it is normally possible to determine which is superior.

(c) There are several possibilities here. If the molecule occurs in a crystal which is a needle and which contains one or more heavy elements, elongation of all ellipsoids in the same direction could occur as a consequence of an inadequate absorption correction (if this is true, most of the other ellipsoids in the structure will look the same). Another possibility is that the molecule is disordered within a lattice and this is currently inadequately represented by only one disorder. It is also possible that the space group is wrong and as a result the molecule is subject to incorrect symmetry. In fact, the ellipsoids are disc-like (oblate) rather than elongated (prolate), a distinction that is hard to make from the view given, and this is attributed to an incipient solid-solid phase transformation in HCN.

(d) The caption could include the following: (a) that the atom numbering system is shown; (b) the probability level used; (c) that the H atoms are drawn as small spheres with arbitrary radii; (d) that the figure shows part of the hydrogen-bonded network. You could add labels denoting the symmetry-related atoms which form contacts to the reference water and organic molecules (e.g. $O9^i$) and to the hydrogens involved. It would then be necessary to define the symmetry operators used to generate $O9^i$ from O9, etc.

(e) Place the bonds involving Cd1 first, followed by those for Cd2, Cd3 and Cd4. Use the same order for the atoms bonded to each Cd (e.g. Cd-O/S/S/Cl, with lower-numbered S atom first). Bonds like Cd4-Cd1 should be inverted to Cd1-Cd4. The Cd-Cd bonds could either be grouped at the beginning or at the end: as they are long, I suggest at the end. A fuller table heading, perhaps identifying the compound involved and the units (presumably Å) used, is required. Align the rightmost column correctly.

(f) Most of the changes will involve reformatting; placing some symbols in italics, subscript or superscript; putting in more elegant, non-ASCII characters; removing redundant text, and replacing CIF keywords with comprehensible text. Although not shown here, the line spacing could be increased (from 1 to 1.5).

Table 1. Crystal data and structure refinement for **5** at 150(2) K

Empirical formula	$C_{24}H_{33}N_3$
Formula weight	363.53
Crystal description	colourless tablet
Crystal size	$0.54 \times 0.50 \times 0.27$ mm
Crystal system	Monoclinic
Space group	$P2_1/c$
Unit cell dimensions	$a = 12.736(6)$ Å $\alpha = 90°$
	$b = 15.371(6)$ Å $\beta = 94.12(8)°$
	$c = 10.676(5)$ Å $\gamma = 90°$
Volume	$2084.6(16)$ Å^3
Reflections for cell refinement	58 measurements at $\pm\omega$
Range in θ	12.5 to 17.5°
Z	4
Density (calculated)	1.158 Mg m^{-3}
Absorption coefficient	0.068 mm^{-1}
$F(000)$	792
Diffractometer type	Enraf Nonius CAD4
Wavelength	0.71073 Å
Scan type	ω/θ

Reflections collected	3680
θ range for data collection	2.65 to 25.03°
Index range	$-14 \to 15, 0 \to 18, 0 \to 12$
Independent reflections	3680
Observed reflections	$2412\,[I > 2\sigma(I)]$
Decay correction	variation $\pm 7\%$
Structure solution by	direct methods
Hydrogen atom location	introduced at calculated positions
Hydrogen atom treatment	riding model
Data/restraints/parameters	3680/0/253 (least-squares on F^2)
Final R indices $[I > 2\sigma(I)]$	$R_1 = 0.0585,\ wR_2 = 0.1304$
Final R indices (all data)	$R_1 = 0.0975,\ wR_2 = 0.1578$
Goodness-of-fit on F^2	1.04
Final $(\Delta/\sigma)_{\max}$	<0.001
Weighting scheme	calc $w = 1/[\sigma^2(F_o^2) + (0.62P)^2]$
	where $P = (F_o^2 + 2F_c^2)/3$
Largest difference peak and trough	0.21 and -0.21 e Å^{-3}

Chapter 16

16.1   The original CIF is shown below, with comments.

⌐—This is not a good start—the CIF was probably prepared in a word processor and it
 ↓ has translated badly—see below.

ü@ü@ü@ü@ü@ü@ü@ü@ü@ü@# ================================
         INITIAL PUBLICATION REQUEST
===

_publ_contact_author
'Prof. John Smith'

;

Faculty of Pharmaceutical Sciences | No. of data items > no. of data names.
John Smith University | Address will be lost.
John Smithsville
IN 40678
U.S.A.

;   ↓————Space not allowed.
_publ_contact_author_email   J. Smith@xray.jsu.edu
_publ_contact_author_fax   '(1) 727 564 2209'
_publ_contact_author_phone   '(1) 727 564 2925'
_publ_contact_letter

;

_publ_requested_coeditor_name   ?
_publ_requested_journal   'Acta Crystallographica, Secti
                                 ↑— Line may have been >80 characters. Information
                                     has been lost and the text string has not been
                                     terminated.

```
_publ_author_name
_publ_author_address
'John Smith'                              # lastname, firstname  ←—minor syntax error
;
Faculty of Pharmaceutical Sciences
John Smith University
John Smithsville
IN 40678
U.S.A.
;
# ================================================
# TEXT
# ================================================
    _publ_section_title
; A new copper (àá) complex of acetic acid
;
_publ_section_abstract
; In the title compound, bis(acetato)copper (àá) dihydrate
```
there exists a three-dimensional network of N–H üc O and O–H üc O
intermolecular hydrogen bonds.

```
; _publ_section_comment
; Acetic acid, (àƒ ) ,and its derivatives have been the subject of recent interest. For
```
example, copper (àá) dihydrate, (àá) has been determined as a model compound
for probing the interaction ....
The crystal structure is stabilized by the N–H üc O and O–H üc O
hydrogen bonds involving the water molecules:
N(3)–H(3) üc O(2^i^)=2.82(5); N(1)–H(1) üc O(3^ii)=2.82(5);
O(3)–H(31) üc O(4^iii^)=2.95);
                                                            ⌐ caret (^) missing
....
;
                                        ▢ represents translation faults.

```
# ================================================
# EXPERIMENTAL
# ================================================

_chemical_compound_source
; The green acicular crystal was obtained by the slow
concentration of an aqueous solution at room temperature.
;                                        ↓ ↓ ↓ ————spaces missing.
_chemical_formula_sum                    'C9 H6D6N6O12 Cu'

_chemical_formula_moiety        Cu(II)[(CH3N3O4)~4~(D2O)~2~]2(D2O)'
```
                                         ↑
                                         ⌐————spaces and commas
                                              missing
                                         ⌐—'~' not allowed here

┌──space should be underscore

_chemical_formula_weight   ' 444.44' ←—quotes are not necessary
_symmetry_cell_setting   'triclinic' ←—quotes are not necessary

_symmetry_space_group_name_H–M   'P1^-^' ←—not standard symbol.
_cell_length_a   5.267(2)
_cell_length_b   7.440(13)
_cell_length_c   9.872(2)
_cell_angle_alpha   86.15(2)
_cell_angle_beta   72.33(2)
_cell_angle_gamma   86.89(2)
_cell_volume   354.3 (4)
                                                    └— illegal space

_cell_formula_units_z   1
_exptl_crystal_density_diffrn   '2.243' ←—quotes are not necessary

_exptl_crystal_density_meas ⌉   ?
_exptl_crystal_density_meas ⌋   'none' ←—quotes are not necessary
....
                                    └──→ data item appears twice in same block.

_computing_data_reduction
                                        ┌—This type of text string cannot
                                        ↓ extend beyond one line.
'TEXSAN. TEXRAY Structure Analysis Package (Molecular Srtucture Corporation,
1985)'
                                                    ↑
                                                    └—typographical error

_publ_section_references
;
Beurskens, P. T. (1984). Technical Report 1984, Crystallography lab. Toernooiveld,
6525 Ed Nijmegen, Netherlands.
....
Sheldrick, G. M. (1986). SHELXS86. Program for the solution of crystal structures,
University of G\"ottingen.
    ←—closing semi-colon (;) is missing.
                                ┌— No. of data items ≠ No. of data names: content of all four
                                ↓   captions will be lost.
_publ_section_figure_captions
'Fig. 1'
; ORTEPII (Johnson, 1976) drawing of the title compound with the atomic
numbering scheme, viewed along the a axis.
Ellipsoids for non-H atoms correspond to 50% probability.
;
'Fig. 2'
; Packing diagram of the title compound along the a axis of the unit cell;
intermolecular hydrogen bonds are represented by dashed lines.
;

┌─ typing error—data name will not be recognised and Table headings
│   will be lost.
↓

_publ_sectiom_table_legends
'Table I'
; Fractional atomic coordinates and equivalent isotropic thermal parameters
(\%A^2^).
;
'Table II'

; Selected bond length (\%A) and angles (^o^).
....   └─should be lengths   └─should use \%
loop_
_geom_bond_atom_site_label_1
_geom_bond_atom_site_label_2
_geom_bond_distance
_geom_bond_site_symmetry_1
_geom_bond_site_symmetry_2
_geom_bond_publ_flag
Cu(1)   O(1)   2.446(8)   .   .   yes
Cu(1)   O(6)   2.468(5)   .   .   yes
Cu(1)   N(5)   1.985(4)   .   .   yes
O(1)   C(2)   1.241(7)   □ ·□ yes   □ represent alternative positions of one
                                            missing place holder (or one symmetry
C(1)   C(2)   1.394(9)   .   .   yes   operator)
....

                    ┌──────────── This should be a comment (preceded by a #) or omitted.
                    │
                    ↓
; Anisotropic_displacement_parameters.
loop_ _atom_site_label

  _atom_site_U~11~ ⎫
  _atom_site_U~22~ ⎪
  _atom_site_U~33~ ⎬   Tilde symbols are not part
  _atom_site_U~12~ ⎪   of these data names—they
  _atom_site_U~13~ ⎪   will not be recognised
  _atom_site_U~23~ ⎭

Cu(1) 0.023(4) 0.021(4) 0.017 (5) −0.007(3) 0.008(4) −0.016(5)
                          └─illegal space.

....
....

Index

Printed in the United States
130064LV00006B/8/A